Bullet Dodging

A Risk Management Handbook for Information, Communication, and Technology (ICT) Projects

Alan Bergsma

Bullet Dodging - A Risk Management Handbook for ICT Projects / Alan Bergsma

Published by Alan Bergsma

First published in 2020

Alan Bergsma, author.

Bullet Dodging - A Risk Management Handbook for ICT Projects: A risk identification handbook for project managers / Alan Bergsma

NATIONAL LIBRARY OF AUSTRALIA

A catalogue record for this work is available from the National Library of Australia

Alan Bergsma at ALB Consulting
www.ictrisk.com and www.albconsulting.com.au
PO Box 1259
Paradise Point,
Queensland, 4216
Australia

Cover design by Thorpe-Bowker, www.myidentifiers.com.au/self-publishing/book-cover-design

Cover image by Thorpe-Bowker

Typeset by Luanne Thibault

The web addresses referenced in this book were live and correct at the time of the book's publication but may be subject to change.

ISBN 978-0-6488700-1-2

Contents

List of Diagrams

List of Tables

Bonus Extras - Tools, Templates, and Examples

Thank you for purchasing *Bullet Dodging - A Risk Management Handbook for ICT Projects*.

There are extra bonus tools, templates, and forms that I provide to readers for free to help you manage risk in your project. These are listed below against the section within the book in which they are explained. Please visit www.ictrisk.com.

Bonus Extra tool, template, or example	Book reference
Project Risk Profile Assessment Tool	Sections 3.3 & 6.2
Pairwise Comparison Tool (AHP)	Section 3.2.1
Reserve Calculator Tool	Section 3.4.2
Risk Register Template	Section 3.4.4
Category Printout Template	Section 4.2.1.12
Probability & Impact Assessment Sheets	Section 5.6.3

Preface

This book is intended to provide you with a practical guide to applying risk management theory to real-life ICT projects. This book is intended to provide an answer to the question I asked over twenty years ago: "How can I best manage delivery risks for my ICT project?"

> *"Life is risky. I'll tell you how risky life is; you're not going to get out alive!"*
>
> Jim Rohn - American entrepreneur, author, and motivational speaker

A life worth living requires taking risks. Living life without risk leads to stagnation and decay.

So too with organizations. Any organization that wants to get ahead, or even maintain their position, needs to take risks. Risks are necessary if we are to achieve our business objectives.

The choice is not simply whether to take a risk or to not take a risk. Our businesses are at risk if we embrace change, but they are also at risk if we avoid change. We are exposed to risk no matter what we do. There is no getting away from risk in business.

> *"Avoiding danger is no safer in the long run than outright exposure. The fearful are caught as often as the bold."*
>
> Helen Keller, Author

As organizations can't avoid risks, organizations choose to proactively take on a risk by initiating change. Organizations initiate change through projects. And, projects always introduce risk of their own.

When we initiate projects, we engage in activities outside of a business' norm. When an organization embarks on a project, they act outside of their normal business-as-usual activities. Acting outside of normal business-as-usual activities is risky by its very nature.

With the increased pace of technology change, the riskiest projects that most organizations embark on are ICT projects, which are notoriously complex. That increased risk results in a high failure rate for ICT projects.

That increased risk translates into a shocking number of projects blowing their budgets. McKinsey's research, conducted in collaboration with the University of Oxford, suggests that half of all large IT projects massively blow their budgets.

According to the study, 17 percent of IT projects go so bad that the projects threaten the very existence of the company. These projects succumbing to unpredicted high-impact risk events which occur significantly more often than would be expected under a normal statistical distribution. Large IT projects that introduce existential risk to their organizations often have budget overruns from 200 percent to 400 percent.[1]

So, why do organizations initiate ICT projects if the failure rates are so high? It is because ICT projects can provide efficiencies that promise disproportionately large benefits.

> *"We take risks knowing that risks will sometimes result in failure, but without the possibility of failure there is no possibility of success."*
>
> Tim Cook - CEO of Apple

The very nature of many ICT projects allows organizations to leverage the efforts of their workers to multiply their productivity. And, ICT projects can also meet regulatory requirements that require information to be stored, managed, and accessed in ways that would be almost impossible without an ICT system.

Organizations need to adopt new technologies just to maintain their current market position, lest their competitors or upstarts steal their market share through new technologies. They need the productivity increases that ICT projects can deliver whilst efficiently meeting regulatory requirements to stay competitive. And, to implement these new technologies, organizations need ICT projects.

> *"You've got to take risks if you're going to succeed."*
>
> Sir Richard Branson - Founder of the Virgin Group

It is the high value of the promised benefits that motivates organizations to initiate ICT projects, despite the high risk. The return on investment (ROI) is high enough that organizations can justify the risk of projects failing.

Fortunately, quantitative risk analysis gives us the tools to assess the likelihood of achieving our target ROI. If we engage in quantitative analysis before we sign off on our business case, we'll be able to make a much more informed investment decision.

Just because ICT projects are inherently risky, that doesn't mean that organizations need simply accept the risk. Just because the benefits outweigh the risks, it doesn't mean organizations don't have a responsibility to manage the risks.

Organizations need to take risks if they are going to realize their objectives. Organizations that manage risks well recognize that they can attain a competitive advantage by actively managing both opportunities and threats. Managing risk is the very center of business survival and growth.

Managing risks effectively will improve performance against project objectives through

- Reduced waste
- Less fraud
- Improved service delivery

[1] https://www.mckinsey.com/business-functions/mckinsey-digital/our-insights/delivering-large-scale-it-projects-on-time-on-budget-and-on-value

- Less management time spent on issues
- More efficient and effective use of resources
- Reduced cost of capital
- Increased innovation
- Fewer unwelcome surprises and sudden shocks
- A higher success rate for change initiatives[2]

For private organizations, these benefits create more shareholder value. And, for public organizations, these benefits lead to better cost performance and improved compliance with legislation and policies.[3]

Risk management best practices help organizations to better understand the risks they face when they embark on ICT projects. And risk management best practices help organizations develop the most appropriate plans on how to manage those risks.

In this book, I aim to explain and expand on internationally accepted best practices with descriptions and examples of how the concepts can be applied in real-world situations.

Employing internationally accepted risk management best practices will help you to

- Gain a better understanding of the risks that your project faces. Working through the risk management process outlined in this book with give you a comprehensive understanding of threats and opportunities.
- Develop proportionate, valuable, and appropriate risk response plans. This book will ensure that your risk response plans are balanced and always provide more benefit than they cost.
- Enable your sponsor / project boards to make better-informed decisions. If you follow the advice in this book and you identify, analyze and report on risks as per the industry best practice, your sponsor & boards will be able to make better informed decisions.
- Enhance your reputation and bolster your career. Whether in the public sector or in the private sector, people in charge of projects are entrusted with the organization's resources. Using internationally accepted risk management best practices provides evidence that you've taken those responsibilities seriously, resulting in larger projects in the future.
- Provide an audit trail of best practices. If your project falls victim to a debilitating risk, your organization may arrange to have an audit performed. If you have managed your risks according to the advice in this book, you will be able to provide the auditors with solid defensible evidence that you have followed risk management best practices.

Given the importance of risk management, it is surprising how little guidance organizations give to their project practitioners.

[2] Axelos M_o_R, pg. 4
[3] Axelos M_o_R, pg. 5

In this book, you will learn the specifics of the entire risk management process covering the full project lifecycle, with specific emphasis on risk identification. This emphasis is important for three reasons.

First, risk identification is the aspect of risk management that I've seen projects apply the least amount of rigor to. Hence, it is often the aspect of risk management where there is the greatest opportunity to improve.

Second, risk identification is the most critical aspect of the risk management process. Risk identification is critical as it feeds the rest of the risk management process. If we miss risks in the risk identification step, there is nothing to refine, analyze, or devise risk responses for and nothing to monitor.

Third, projects that fail, usually fail, because they didn't identify a risk. It is far less often that projects fail because they've identified a risk but managed that risk poorly.

Risk identification is so critical that we provide a framework for risk workshops in this book. These risk workshops will illustrate why and how risk identification tools and techniques can produce the most value.

The downside of risk workshops is that they can degenerate into a waste of time and cause a loss of face if we don't manage them well. It is not easy to run a risk workshop effectively. It needs to be well planned, well-structured, and well facilitated.

Over two decades ago, I was asked to facilitate my first risk workshop for a project. I had participated in a couple of risk workshops, but they were unstructured gab sessions that achieved little. They were not examples that I wanted to follow. So, I started researching the subject. Over 20 years of research has resulted in this handbook containing a one-stop shop offering clear advice on how to facilitate a risk workshop.

Thanks to the information I found, the workshop I facilitated was well received by the project stakeholders. Since then, I have had the opportunity to facilitate many more.

Fortunately, the project risk management industry has matured since those early days. PMI has developed the Risk Management Professional (RMP) accreditation. And Axelos has developed the M_o_R accreditation. These accreditations have lifted the importance and focus of risk management within the project management discipline. Both accreditations adhere to the internationally recognized ISO:31000 - risk management standards.

The PMI and Axelos methods of risk management vary, but their approaches are complimentary rather than contradictory. I draw from both of their approaches in this book.

Both the PMI and Axelos approaches provide ICT projects with tools and techniques to manage risks. They both recommend *what* to do to manage risks. But neither go into detail on *how* to apply those tools and techniques in risk management.

That is why, in this book, I expand and build on the excellent risk management work done by PMI and Axelos. In this book, I provide explanation and examples on *how* the recommended risk management tools and techniques can be used in ICT projects. I also provide a broader context, to help make conducting risk management easier.

This book provides the next level of detail to complement the PMI and Axelos publications. Our goal is to provide a "how to" guide to help people apply risk management tools and techniques to ICT project risk management. This book is meant to provide you with the guidance on risk management that I was looking for decades ago.

Consider using a combination of tools and techniques, because no single tool or technique can be expected to identify all risks. All tools and techniques have their own strengths and weaknesses. Please feel free to pick and choose tools and techniques in this book to create a combination that is appropriate for your project.

This book is meant to be a practical guide and not mired in theory. This book is not intended to be academic, but instead to serve as a handbook that you can use as a reference guide, wherever you are in your project lifecycle.

Organizations face many different types of risks. Many of these are corporate risks that fit into BAU processes that have been designed to manage them.

Project risks are different. Each project is unique, and the risk management requirements are different for each project. Those unique requirements of projects are why I have focused this book on projects.

A few of the many different types of risks that organizations face can be seen in the table below. The reddish column denotes external risks, the green column denotes enterprise and operational risks, and the blue column represent special operations risks. Project risks are a type of special operations risks.

When we talk about project risk, we are talking about the risk of not delivering the project according to the expected project objectives: scope, quality, time, and cost.

Table 1 – Program and project risks contextualized within the broader types of business risk

Special Ops	External	Enterprise & Ops
Program Risk	Competitive Risk	Reputational Risk
Project Risk	Interest Rate Risk	Process Risk
Innovation Risk	Economic Risk	Strategy Risk
	Credit Risk	Compliance Risk
	Legal Risk	Operational Risk
	Exchange Rate Risk	Quality Risk
	Resource Risk	

There are many aspects of generic project risks across industries. But, in this book, I have focused on ICT projects because ICT projects have unique risk considerations.

All of my 20+ years' experience has been on ICT projects. My project experience spans several industries, but they have all been ICT projects.

Whether the projects I have delivered have been in financial services, government, natural resources, or in the intelligence community, they have all been ICT projects. And ICT projects have their own distinct risk management requirements.

It is worth clarifying that this book is about ICT project delivery risk. This book is not about ICT security risk or cybersecurity risk, other than as they pertain to ICT project delivery risk. This book is about the risk of initiating an ICT project, and then failing to deliver what was promised.

Many authors shy away from the use of examples in their books on risk management. I can understand why. The use of examples exposes the author to criticism. It is much safer to stick to theory. But, examples provide significant value to the reader by bringing the tools and techniques to life. Examples allow the users to see how the tools and techniques can apply to real-life situations.

Is your organization about to launch a significant ICT project? Or, has your organization already started a significant ICT project? Whether you're a project manager, a member of the project board, or an executive, you need to know how to manage risks. In an ICT project, you need to know how you are going to manage risk to deliver value and to avoid failure.

I wrote this book to make project risk management easier and more effective for ICT projects. Whatever your role in the ICT project, I have written this book to help you manage risks and to achieve your project objectives.

I have written this book to help you manage risks for your project in a way that is realistic and leads to your project success.

To be effective, we need to accept the fact that risk exists. We need to be realistic, and we need to plan to build the future that we want. Simply hoping that success will occur will not give us the future that we desire. We need to recognize that if we try to avoid risk entirely, we will decrease our chances of achieving our goals.

Let us plan to build the future that we desire. Let us accept that risk exists. And, let us go forward with our eyes wide-open to the uncertainty our projects face. Let us use the best practice tools and techniques of project risk management to maximize our chances of achieving our project goals.

> *"The price of inaction is far greater than the cost of making a mistake."*
>
> Meister Eckhart, Theologian, philosopher, and mystic

Let us plan and act in our projects with our eyes wide-open to the risks involved.

If you are experienced in risk management and have successfully developed your own approach and techniques, you may not need all the tools and techniques I describe in this book. But I hope there are at least some tools and techniques in this book that you can benefit from. Either way, I wish you success in your risk management and in your project delivery!

PS: If you're curious to find out how your project's inherent risk profile compares to industry benchmarks, why not take our free five-minute *Project Risk Profile Assessment*? It will give you an evaluation of your project's risk profile against industry benchmarks. Our *Project Risk Profile Assessment* tool is available for free for anyone who has purchased this book.

Our *Project Risk Profile Assessment* tool generates a printable report that you can take away for your consideration or to share. Please visit www.ictrisk.com for your free project assessment.

Glossary of Terms

As with many industries, risk management has developed a lexicon of its own. In the table below, I have outlined the specific meaning of the terms as they are used in risk management. I have tried to keep the definitions congruent with international standards but, in the absence of standard definitions, I have used subjective descriptions.

Table 2 – Glossary

Term	Meaning
Consequence	Outcome of an event affecting objectives (synonymous with "impact"); a consequence can be certain or uncertain and can have positive or negative effects on project objectives. Those effects can be either direct or indirect
Contingency reserve	Time or money allocated in the budget or schedule baseline for known risks with risk response plans
Emergent risk	Emergent risks are risks that arise after the initial risk identification. These are risks that the project either failed to identify or that have emerged due to either an internal or external change. Emergent risks are typically simply added to the risk register under "inherent risks" and then are analyzed as the original inherent risks
Event	Occurrence or change of a particular set of circumstances; an event can also be something that is expected which does not happen, or something that is not expected which does happen
Expected risk	The risk level that the team expects to achieve after the risk response plan has been implemented
Impact	Synonym with *consequence*; see *Consequence*
Inherent risk	Inherent risk is the risk that exists if no controls or other mitigating factors were put in place (the gross risk or the risk before risk response plans are implemented); this is the risk score for the probability and impact before you've done anything to affect them
Likelihood	Chance of an event occurring.

Term	Meaning
Management reserve	An amount of the project budget or project schedule held outside the performance measurement baseline for management control purposes that is reserved for unforeseen work that is within the project scope
Opportunity	An uncertain event that would have a positive impact on objectives if it occurred
Probability	Synonym with *likelihood*; see *Likelihood*
Residual risk	Residual risk is the remaining risk after you've implemented a risk response plan. It's the risk remaining after you've reduced the risk, removed the source of the risk, modified the consequences, changed the probabilities, transferred the risk, or retained the risk.[4] For example, if your risk response plan were to mitigate a risk, your residual risk will be the risk that is left over after you have mitigated the risk by reducing the probability and/or reducing the impact.
Risk	The effects of uncertainty on project objectives; usually expressed in terms of *cause*, potential *events*, or their *effect*
Risk appetite	The degree of uncertainty an organization or individual is willing to accept in anticipation of a reward "The amount and type of risk that an organization is prepared to pursue, retain, or take in pursuit of its objectives"[5]
Risk attitude	Risk attitude is "An organization's common approach to risk. An organization's *risk attitude* guides how risks are evaluated and managed. An organization's attitude towards risk determines whether risks are taken, tolerated, retained, shared, reduced, or avoided, and how risk response plans are implemented."[6]
Risk exposure	The combined effect of risks to a set of project objectives; the extent of the risk borne by a project at a given time
Risk management	Plan, identify, coordinate, direct, and control a project's uncertainty
Risk proximity	The amount of time between the present and when the risk could possibly occur.
Risk response plan	The actions that will be taken to treat a risk. Risk response plans can include accept, escalate, avoid, transfer, mitigate, enhance, exploit, or share
Risk source	Element which alone, or in combination, has the potential to give rise to risk

[4] http://praxiom.com/iso-31000-terms.htm
[5] ISO 31000:2018 standard on risk Management
[6] https://www.praxiom.com/iso-31000-terms.htm#RiskAttitude

Term	Meaning
Risk threshold	The measure of acceptable variation around an objective that reflects the risk appetite of the organization
Risk tolerance	The degree of uncertainty that an organization or individual is willing to withstand
Risk treatment plan	Synonym with *risk response plan*; see *Risk response plan*
Secondary risk	Secondary risk is a risk that arises as a result of the risk response plan that you enact for the inherent risk. For example, if your risk response plan for a risk is transfer by engaging a third-party vendor to do a portion of work, you are introducing a secondary risk that the third-party vendor may fail to meet their contractual obligations
Stakeholder	Person or organization that can affect, be affected by, or perceive themselves to be affected by a project decision or activity; also, sometimes called "interested party"
Threat	An uncertain event that would have a negative impact on objectives if it occurred
Trigger	Events or conditions that indicate a risk is about to occur
Watch list	A list of risks that either have a low-priority or require further analysis. Risks in a watch list may not need proactive management action beyond being placed in the risk register or adding a contingency reserve
Work breakdown structure (WBS)	A tiered breakdown of the total scope of tasks to be carried out by the project team to complete the project objectives and create the required deliverables

"Risk is like fire:
if controlled, it will help you;
if uncontrolled, it will rise up and destroy you."

Theodore Roosevelt,
26th president of the United States

1 Introduction

In today's society, people are far more aware of risks than they were twenty, or even ten, years ago. Technology has accelerated communication and response times. A risk realized in a remote corner of the globe can impact an organization immediately. Gone are the days of pushing back decisions on how to deal with change until next month or even next week.

Today's business environment is risky. Implementing change is risky. Maintaining the status quo is risky. There is no hiding from risk in business today.

> *"The biggest risk is not taking any risk. ... In a world that's changing really quickly, the only strategy that is guaranteed to fail is not taking risks."*
>
> Mark Zuckerberg, CEO Facebook

And, ICT projects, especially large ones, constitute a disproportionately large portion of the risks businesses face.

McKinsey's research, conducted in collaboration with the University of Oxford, suggests that, on average, large IT projects run 45 percent over budget and 7 percent over schedule, while delivering 56 percent less benefit than predicted.[7]

Consider what that would mean for a hypothetical $15,000,000 project that is to be delivered over 24 months and deliver benefits worth $25,000,000.

As we can see in the below table, 45% over on a budget of $15 million creates a budget overspend of $6.75 million. The 7% schedule blow out means the project will take just over 7 extra weeks to deliver. And, if we assume the benefits are financial, the under-delivery of benefits by 56% means that we have a benefits shortfall of $14 million!

The business case of our hypothetical ICT project was approved based on a healthy ROI of 67% or $10 million in net gain. But, if the research findings hold true, our hypothetical project is going to actually deliver a negative ROI of 56% or a $10.75 million net loss.

	Promised	Actual	Shortfall
Cost	$15 million	$21.75 million (45% over)	- $6.75 million
Schedule	24 months	25.68 months (7% over)	- 7 weeks
Benefits	$25 million	$11.00 million (56% under)	- $14 million
ROI	67%	- 56%	

[7] https://www.mckinsey.com/business-functions/mckinsey-digital/our-insights/delivering-large-scale-it-projects-on-time-on-budget-and-on-value

If we extrapolate the research findings on our hypothetical project further, we find that our hypothetical project which started out promising to deliver benefits of $10 million more than it cost, actually ended up costing $10.75 million more than the benefits actually delivered! That means that the project delivered of $20.7 million less benefit than it originally promised! And that on a project with an initial budget of $15 million!

If the abysmal project delivery results from McKinsey's research hold true, our $15 million hypothetical project would need to have originally promised benefits of $49.5 million, just for the project to break even!

Clearly, delivering ICT projects 45% over budget, 7% over schedule and with 56% less benefits than promised is not acceptable. And that's where risk management comes in. Managing risks according to the industry best practices outlined in this book will shield your project from such outrageous failures.

Research like McKinsey's brings to the fore the need for well thought through and structured risk management framework. We need to utilize a structured framework to enable organizations to frontload much of the planning and effort required to deal with change, and to do so before the change occurs.

Fortunately, most organizations recognize this. They implement an enterprise risk capability. They implement a financial risk capability. And they incorporate a risk management process into their project delivery framework and project governance.

And, it is the project risk management capability that is particularly interesting. Project risk management requirements are unique and distinct from an organizations' other risk management requirements. Project risk management focuses on the risk of delivering the project according to the project objectives that have been set for schedule, cost, quality, and scope.

There is no aspect of a project that is imune from risk. That is why project risk management spans the project lifecycle. Project risk management starts from the initial project planning activities and goes all the way through to project closure.

It is worth noting here the importance of performing aspects of quantitative risk analysis before the business case is finalised. When deciding to approve the funds for an ICT project, aspects of quantitative risk management can provide insight into the project that will influence investment decsions.

Specifically, performing Monte Carlo analysis before finalising the business case will provide invaluable information to the investment decision. Monte Carlo analysis can help the decision makers understand the probability of achieving the claimed ROI, the probability of achieving a negative ROI, the amount of contingency needed to achieve a set probability of success and the factors most influencing the project's risk profile.

Monte Carlo analysis is especially valuable before the business case is finalized, but there is much more to project risk management. At the heart of the project risk management is risk identification. Risk identification provides the raw materials that the rest of the risk management process uses. The output of the risk identification process is the input to all the subsequent risk management activities.

If you don't identify risks well, it doesn't matter how well you execute the rest of the risk management process.

Risk Identification is the input that feeds the rest of the risk management process. If we fail to identify risks or we fail to identify risks correctly, the analysis of risk, treatment of risk and monitoring of risk will be of little consequence.

That being said, this book covers the whole of risk management for ICT projects. It provides tools and techniques for managing the process throughout the project lifecycle.

This book is intended to be a resource that you can reference at any point in a project and has been structured so that each section can be read in isolation. The content of each section doesn't require the reader to recall information from other sections. In order to achieve that, there is some repetition in this book. If you are reading it from start to finish, please forgive any repetition knowing that it will help you when using the book as a reference.

The PMBOK 6th edition lists seven risk management processes. This book covers all seven processes but emphasizes the following four processes, identify risks process, qualitative risk analysis process, quantitative risk analysis process and plan risk responses process.

The diagram below shows which of the PMBOK processes are emphasized in this book.

Diagram 1 - Primary Focus Diagram

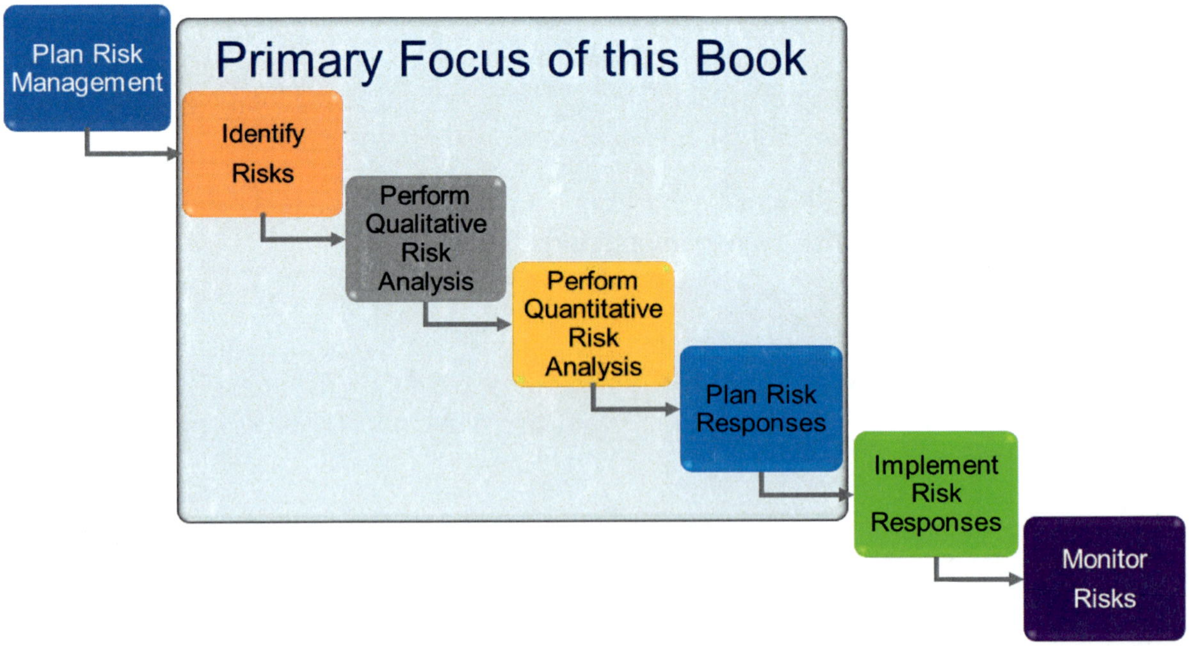

This book starts out with the plan risk management process. Without a plan, the risk management process would be an unstructured mess. Without a plan, that unstructured mess could not effectively manage the risks of a project. Without a plan, a project's chance of success would drop dramatically.

After the plan risk management stage, risk identification is the most important stage of the risk management process. Risk identification is the stage of the process where there is the greatest opportunity for improvement.

Project managers (PMs) typically identify risks by conducting a risk workshop with the project stakeholders. Gathering together a diverse group of disparate stakeholders enables the project to draw from a broad spectrum of expertise and experience.

The participants often view these workshops as onerous, but they need not be. Properly managed, these workshops can be engaging and thought provoking for your stakeholders. They can provide participants with tools and techniques useful in other aspects of their day-to-day jobs.

But, your participants are not the main beneficiaries from risk identification workshops. The PM reaps the most benefits from a risk identification workshop.

A risk workshop gives a PM a chance to thoroughly examine the project's uncertainties. It gives the PM a chance to hear about the project uncertainty from a variety of disparate viewpoints.

These varied and distinct viewpoints can help identify uncertainties that might otherwise cause a project to fail. Whether the PM facilitates the workshop, or if the PM has a third-party facilitator, the PM bears responsibility for the workshop's success.

If a PM facilitates the workshop, the risk workshop gives a PM the opportunity to differentiate themselves from their peers.

PMs rarely conduct risk workshops more than a few times a year. As such, they rarely take the time and effort to develop the knowledge and skills to run a risk workshop well.

But, by using the tools and techniques in this book, you can differentiate yourself from your peers. You can showcase your understanding of risk management best practices.

The opposite is also true. I've watched some PMs conduct risk workshops that were haphazard and sloppy. These poorly conducted workshops left the door wide open to criticism when risks were later realized. The sloppy workshops damaged the PM's reputation and left the PM susceptible to blame when risks were later realized.

At the heart of the problem is the infrequency that we conduct risk workshops. Most projects conduct risk identification workshops only a few times per year. As such, most PMs don't dedicate the time to gaining the knowledge and developing the skill to optimize their workshops. They spend such a small percentage of their time on risk workshops that they rarely see the value in investing the effort on improving.

That's great news for you! It creates an opportunity for you to differentiate yourself from the average PM. You can differentiate yourself by either using the tools and techniques in this book or by engaging a professional third-party project risk specialist to help.

When risk workshops are well conducted, and all possible areas of risk are collectively explored, the participants take ownership of the project's risks. When your participants take ownership of project risks, you will have a much easier time in managing those risks throughout the life of the project.

I encourage you to use the tools and techniques described in this book in your workshop. Use them to probe and explore every avenue where project uncertainties lurk. Then ask your participants to confirm the list of risks afterwards. If they confirm that the list of risks is complete, then they will share ownership of the risks.

Later in the project, if something detrimental occurs that wasn't identified as a risk in your workshop, you will have fewer fingers pointing in your direction. Well-run risk workshops raise your standing with your stakeholders. They ensure that you have a defensible position should anything go wrong with your project.

Over the last couple of decades, I have worked in project risk management in many different organizations across a variety of industries. In those engagements, it has surprised me to discover how little guidance material I have found available to PMs tasked with identifying risks.

Even in mature organizations with well-developed risk frameworks, organizations provide minimal support material on how to best identify risks.

Yet, the fact remains, if we don't get the risk identification step right, the quality of the rest of the process will be of little relevance. The output of risk identification is the input for the rest of the risk management process. If risk identification is poor, the rest of the risk management process will be poor— "garbage in, garbage out" as the old adage says.

Could it be that the subjective nature of identifying risk is a barrier to documenting risk identification methodology?

Risk identification is a more subjective part of the risk management process than the other, more objective parts of the process. And the subjective nature of risk identification makes it more difficult to describe than other elements of project risk management.

It is possible to apply parameter driven assessments to later tasks in the risk management process. Tasks, such as assessing risk impact & probability and calculating risk scores, are more objective and easier to describe than risk identification.

It is far easier to analyze risks than it is to describe the human behavior required for risk identification. Numbers and equations are much more predictable and manageable than the blue-sky thinking required for risk identification.

Risk identification is subjective and depends on the human element. Could stakeholder turnover disrupt your project? Could your resources be taken away to deal with unforeseen problems? Could a legal or regulatory change impact your project? These are all questions that need the expert judgement of subject matter experts.

These subjective questions cannot be determined through objective methods. You must rely on the expert judgement of your workshop participants. But those expert judgements may not be given readily. You will need to guide, probe, and cajole your participants to get the most benefit from their expert knowledge and experience.

Bringing together a large and diverse group of stakeholders creates a complicated environment. Managing stakeholders with varying interests and diverse opinions exposes the workshop to the complexity and inherent unpredictability of the human element.

Subjective risk identification is a more difficult topic to address than more objective tasks, such as calculating a risk score. But, the strategies, tools, and techniques outlined in this book will help to make risk identification simpler, easier, and more thorough.

A thorough risk identification will feed into the rest of the risk management process. Once you have the risks thoroughly identified, the rest of the risk management process is about competently executing the tools and techniques of the later risk management stages.

> *"In poorly run projects, problems can go undetected until the project fails. It's like the drip ... drip ... drip of a leaky underground pipe. Money is being lost, but you don't see it until there is an explosion."*
>
> Joy Gumz - Director of Project Auditors

In this book, you'll find all the information to guide you through the entire project risk management process from plan risk management to the monitor and control. In this book, you'll find all the tools and techniques required to mend the underground pipe, stem the leak of money, and avoid those explosions.

If you're curious to find out how your project's inherent risk profile compares to industry benchmarks, why not take our free five-minute *Project Risk Profile Assessment*? It will give you an evaluation of your project's risk profile against industry benchmarks. Our *Project Risk Profile Assessment* tool is available for free for anyone who has purchased this book.

Our *Project Risk Profile Assessment* tool generates a printable report that you can take away for your consideration or to share. Please visit www.ictrisk.com.

1.1 Project Delivery Method: Waterfall vs. Agile

With the growing use of Agile delivery methods, some experts are moving away from risk management. They say that the Agile delivery methods reduce risk through the iterative nature of Agile development. While I agree that this may be true of some *product* delivery risks, it is not true of all types of *project* delivery risks.

Risks related to product delivery may be lower when using Agile, but other risks are just as dangerous as if you're using Waterfall methodology. And, Agile methodology introduces some new risks of its own.

Agile methodology can reduce some product delivery risk. Agile methodology provides us with frequent iterations and continuous review cycles that reduce some product delivery risks. The frequent reviews with the user representatives reduce the risk that the delivery team will develop features that the users don't like.

But Agile methodology doesn't reduce all types of risks. Some risks are independent of the product development methodology that your project is using.

Risks related to risk categories such as executive support, communication, commercials, external risks, stakeholders, authority, and environment can be equally dangerous, regardless of whether you're using Waterfall or Agile delivery.

The use of Agile even introduces some risks of its own. For example, security risks are one category that is often highlighted as an area of increased uncertainty when using an Agile delivery method. Frequent releases of software iterations into a production environment makes it difficult to apply adequate security measures.

Relying solely on the nature of Agile methodology to reduce your project risks is kind of like driving a car while only looking at the dashboard. Sure, looking at the dashboard will allow you to see the speedometer, which will enable you keep to the speed limit. And looking at the dashboard will allow you to detect any engine warning lights that may turn on.

But, focusing solely on the dashboard has some obvious limitations. It will not help you keep the car on the road. It will not help you take evasive action if someone is on a collision course. And, it will not keep you from slamming into a car parked on the side of the road!

So too with projects that rely solely on the Agile delivery method to manages risks. They will expose themselves to other types of project risks. If you're using Agile, also use risk management best practices to consider the project environment beyond the development team. Look up from the dashboard to make sure you keep the car on the road, avoid any obstacles and prevent anything else crashing into your project!

I would encourage anyone using Agile methodology to follow risk management best practices outlined in this book. These practices will ensure that risks outside of product delivery don't impact your project.

1.2 Risk Management Principles

I have written this book in accordance with the ISO 31000:2018 risk management principles. These principles are high-level and universally applicable. The principles provide guidance to projects as they help us design appropriate approaches to risk management. The principles, paraphrased below, state that risk management should be

- Integrated – Risk management is a fundamental part of all of an organization's activities
- Structured and comprehensive – A structured and comprehensive approach to risk management helps deliver consistent results; it allows us to compare apples to apples
- Customized – The risk management framework and process are customized to the organization's objectives; it takes into consideration both the internal and external environments
- Inclusive – Involving stakeholders at the most appropriate time and to the most appropriate degree enables the project to benefit from their expertise; involving stakeholders in the risk management process increases their involvement, buy-in, and strategic alignment

- Dynamic – Risks can disappear; risks can change, and new risks can emerge; risk management enables us to predict, detect, and react to such changes and events in a fitting and suitable manner
- Best available information – We conduct risk management on the best information available; that information comes from records of what has occurred in the past, measurements and assessments of what is happening currently, and our expectations of what will happen in the future; information should be timely, well-defined, and accessible to pertinent stakeholders
- Human and cultural factors – Human behavior and culture will impact risk management in every facet; risk management recognizes that human behaviors are inherently unpredictable
- Support continual improvement – Risk management should be improved continually throughout the project, as we develop learnings from our experiences[8]

1.3 Project Constraints

Prince2 lists six performance targets (also referred to as the six tolerances[9]): time, cost, quality, scope, benefits, and risk.

PMBOK lists projects' constraints include scope, schedule, cost, quality, resources, and risk.[10]

As this book is about risk, when referring to project constraints, I use a combination of the Prince 2 and the PMBOK project objectives, excluding the constraint "risk." This is because the performance target of risk doesn't fit well in the risk management process (e.g., "What is the sponsor's risk appetite for the *risk* constraint?" AND "What risks are there in the *risk* constraint?").[11]

In this book, I use a combination of PMBOK and Prince 2 when I talk about project constraints. I use *time, cost, quality, scope, benefits,* and *resources* when referring to project constraints.

Diagram 2 – Project Constraints as used in this book

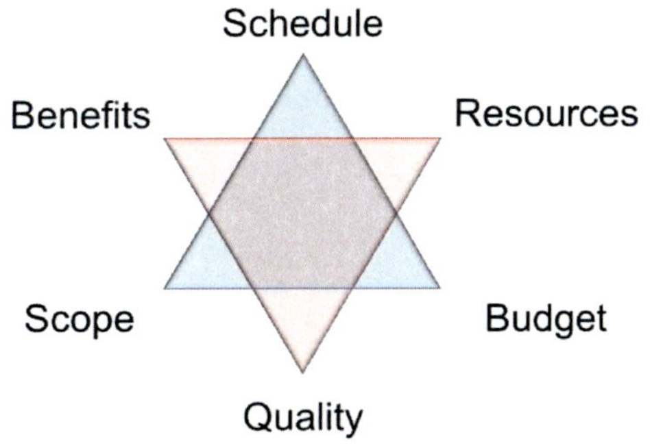

[8] Section 4 of the ISO 31000:2018 Standard
[9] https://www.pmi.org/learning/library/six-constraints-enhanced-model-project-control-7294
[10] PMBOK Part 2 The Standard for Project Management, section 1.1 pg. 542
[11] https://www.pmi.org/learning/library/six-constraints-enhanced-model-project-control-7294

2 ICT Project Risk Management Overview

Risk management in ICT projects is complicated. And, when things are complicated, the simplest and most reliable way to ensure you are thorough is to use a structured approach.

2.1 The Structure of Risk Management

Both Axelos and PMI provide structured risk management frameworks. These frameworks are both similar and complementary. They enable us to take a methodical approach to managing project risks.

This book is structured according to the seven PMBOK risk processes, as they provide a logical framework for risk management that is simple to work with. Those seven processes are

1. Plan risk management
2. Identify risks
3. Perform qualitative risk analysis
4. Perform quantitative risk analysis
5. Plan risk responses
6. Implement risk responses
7. Monitor and control

These risk management process steps are both sequential and cyclical.

2.2 The Sequential Nature of Risk Management

The diagram below shows the chain of sequential process steps. The process steps are performed one after another. From start to finish, from link to link, there is a linear and sequential progress from one step to the next.

Diagram 3 - The Sequential Nature of Risk Management

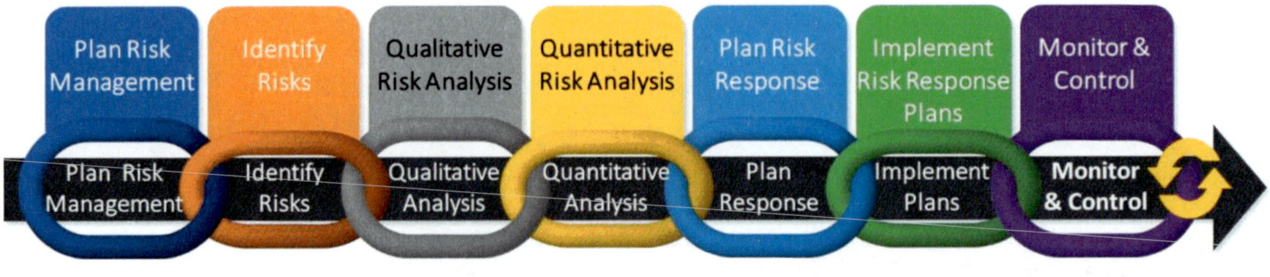

But, the risk management process is not purely sequential. There are concepts in the sequential steps that justify repeating. There are concepts that should be considered in more than one place and at more than one time.

That is why any repetition in this book is intentional. There are some steps in the process that merit reiterating earlier points. Considering the same concept in a different context often allows us to examine distinct and varied aspects of the same concept.

2.3 The Cyclical Nature of Risk Management

There are three aspects of the risk management process that are cyclical. The first is monitor and control, the second is progressive discovery, and the third is incorporating risk management learnings.

2.3.1 Monitor and Control

First, the monitor and control process step is cyclical in its nature. The dynamic nature of projects requires that we continually (iteratively) monitor project risk. In the diagram below, the circular arrows at the end of the monitor and control process step represent this cyclical nature of the monitor and control process step.

Diagram 4 - Cyclical Nature of the Monitor and Control Process Step

The monitor and control step involves continually checking on the project risks and project risk related actions to ensure that no change has occurred that requires action.

2.3.2 Progressive Discovery

Second, projects undergo progressive discovery throughout the project lifecycle. This progressive discovery necessitates cyclical risk management. Progressive discovery is similar to the project management concept of progressive elaboration, but it is a more active process.

Progressive discovery is when new information is unearthed, and additional details are uncovered. This new information might pertain to new risks or it might change existing risks. Because of this progressive discovery, risk identification, risk analysis, plan risk response, and implement risk response need to be periodically reiterated.

As the project moves through the project lifecycle, new risks will emerge, and existing risks will change, become realized, or become irrelevant. It is due to this progressive discovery that we need to continually cycle through elements of the risk management process.

Risk identification sessions need to be conducted iteratively throughout the project to ensure that areas of uncertainty are probed, and emergent risks are captured. Any emergent risks need to go through risk analysis, plan risk response, and implement risk response steps.

Previously identified risks need to be periodically reassessed to ensure that they are still relevant and to identify any changes. Each existing risk should be assessed and, if necessary, put through the processes of risk analysis, plan risk response, and implement risk response.

The cyclical nature of progressive discovery in risk management can be seen in the diagram below.

Diagram 5 - Cyclical Nature of Progressive Discovery

As you can see in the above diagram, both new emergent risks and existing risks feed into the analyze risks step. This step is followed by plan risk response, implement risk response, and, finally, monitor and control. Risks remain in monitor and control until they circulate back into the process in the next risk review.

2.3.3 Risk Management Learnings

Third, is the cyclical nature of incorporating risk management learnings. Some aspects of the monitor and control process (e.g., risk audit) provide feedback on the effectiveness of the manage risk process. We can apply this feedback as learnings and adjust and amend any of the process steps as necessary.

A risk audit may identify aspects of the risk management plan that are not serving the project effectively. For example, the frequency of a risk management activity could be found to be insufficient. You may decide to update the risk management plan to reflect an increase in the frequency of that risk management activity.

Similarly, a risk audit may identify an aspect of your risk identification process that has not been working well for the project. As a result of detecting this deficiency, you may decide to change the way that you identify risks. For example, you may find that there are some tools or techniques that are detracting from your risk identification sessions, and you may decide to cut them out. Depending on the level of detail of the change, you may also need to update the risk management plan.

A monitor activity may also find that there are aspects of your analysis processes that are not working well for your project. For example, you might find that the format that you use to report the output of your qualitative analysis is confusing for your stakeholders. You may decide to change it. Depending on the level of detail of the change, you may also need to update the risk management plan.

You might find that a monitor activity identifies some aspect of the plan risk response process that is not working for your project. For example, you might find that one of the risk responses that you've been using is disliked by your sponsor. You may decide to stop using it. Depending on the level of detail of the change, you may also need to update the risk management plan.

A monitor activity may find that there are aspects of your implement risk response process that aren't working for your project. For example, you might find that some aspect of how you've implemented one of your risk response plans has not worked and needs to be changed. Depending on the level of detail of the change, you may also need to update the risk management plan.

And finally, a monitor activity may find that there are aspects of your monitor and control process step that need improving. For example, you might find that the risk audits need to be executed more frequently to keep abreast of changes. You may decide to change the frequency of the risk audits. Depending on the level of detail of the change, you may also need to update the risk management plan.

Diagram 6 - Cyclical Nature of Risk Management Learnings

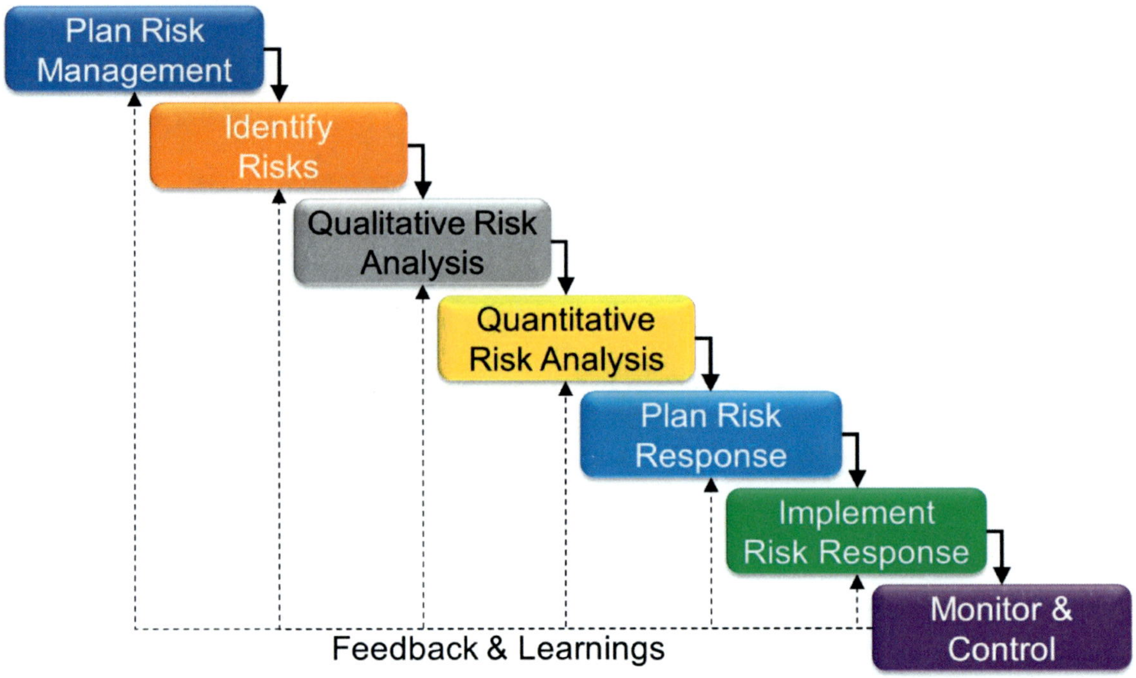

As you can see in the diagram above, the monitor and control step provides feedback and learnings that can be incorporated into any of the earlier process steps. It is that feedback and learnings that makes this aspect of the risk management process cyclical.

2.4 How Much to Invest in Project Risk Management?

It is difficult to provide guidance on how much time and effort should be invested in risk management. The amount you invest will be dependent on your sponsor's risk appetite. That being said, there is a growing trend in the industry to allow 2% of the total project budget on second-line and third-line risk management.[12]

First-line risk management involves the day-to-day management and control activities. These include implementing project controls such as plans, schedules, and budgets, to name a few. First-line risk management activities are often so tightly integrated with other project activities that the risk management component is difficult to segregate. As such, they are not included in the 2% estimate for risk activities. First-line risk management costs are above and beyond the 2% estimate.

Second-line risk management involves defining, enforcing, and maintaining the standards and processes for project risk management. It includes activities that support the project by providing tools to manage new and emerging project risks, aggregating risks, and monitoring the effectiveness of project controls.

[12] Oakes, G. 2008, 'Project Reviews, Assurance and Governance'

Third-line risk management involves independent assurance. It includes independent assessment of the effectiveness of project risk management to provide the project board with confidence that the first and second lines of defense against risk are operating effectively.[13]

If your project is strategically important, involves external partners, is complex, or is large in scope, duration or impact, then you may decide to invest more than 2%.

If your project is not strategically important, doesn't involve external partners, is simple and straightforward, or is small in scope, duration or impact, then you may decide to invest less than 2%.

In the preface and introduction, we talked about some of the shockingly poor results that were uncovered by the research conducted by McKinsey in conjunction with Oxford University.

This research was conducted on a sample size of more than 5400 ICT projects and it found that half of all large IT projects, projects with initial price tags exceeding $15 million, massively blow their budgets. On average, large IT projects run 45% over budget and 7% over time, while delivering 56% less value than predicted.[14]

When you are thinking about how much to spend on risk management, consider what the impact would be of a 46% overspend on your project. Or the impact on your project of delivering 56% less benefits than promised.

I think you'll agree with me, in the light of such risks, spending, 2, 3 or even 4% of your project budget on risk management seems paltry in comparison.

Most projects and programs I have worked with have said that, dollar for dollar, the funds that they invested in risk management was the best value of any project activity.

It is not just the research out of McKinsey and Oxford university that paints such a dismal picture for large ICT projects. The Standish Group's Chaos Report[15] portrays a similar view of large ICT projects. And you only need ask around of a few seasoned project professionals to hear similar stories.

Consider the vast potential downside when you budget for your project's risk management.

2.5 Chapter Summary – ICT Project Risk Management Overview

In this chapter we've gone through an overview of project risk management. We've included a high-level description of

- The structure of risk management
- The sequential nature of risk management

[13] Deloitte. 2017, Project Risk Management Applying the Three Lines of Defence Model to Project Risk Management
[14] https://www.mckinsey.com/business-functions/mckinsey-digital/our-insights/delivering-large-scale-it-projects-on-time-on-budget-and-on-value
[15] https://www.standishgroup.com/sample_research_files/CHAOSReport2015-Final.pdf

- The cyclical nature of risk management including
    - Monitor and control
    - Progressive discovery
    - Risk management learnings
- Guidelines on how much time and treasure to invest in risk management

3 Plan Risk Management

Plan risk management is the first link in our risk management process chain of steps.

Diagram 7 – Plan Risk Management

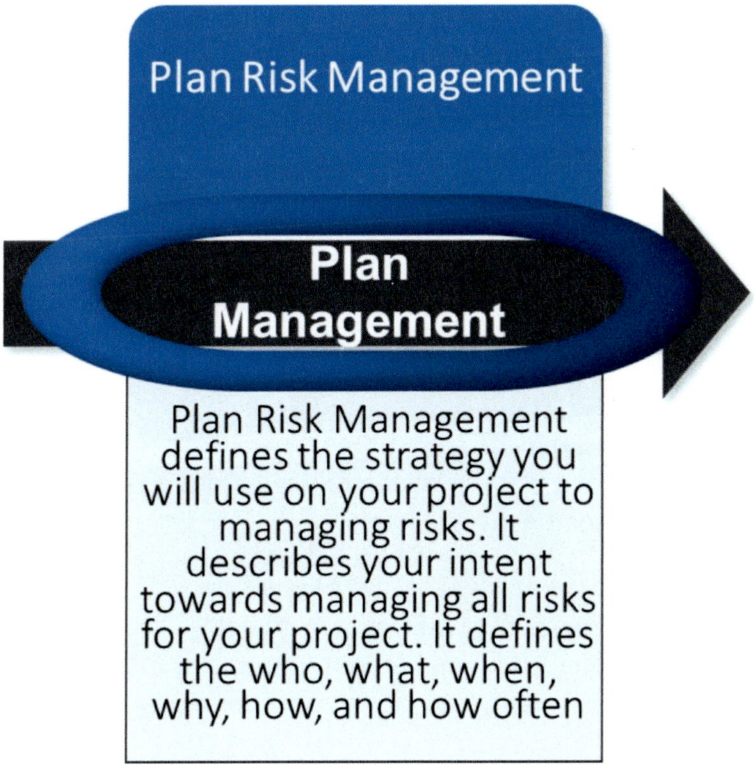

> *"It is only by being bold that you get anywhere. If you are a risk-taker, then the art is to plan to protect the downside."*
>
> Sir Richard Branson, Founder of the Virgin Group

When Richard Branson decided to go into the airline business, he felt that if Virgin could create a really fun, high quality airline that got all the details right, they'd have a chance to survive and even flourish.

So, once he had negotiated the price for a second hand 747 from Boeing, he said to Boeing that if Virgin Atlantic wasn't successful, then he wanted to be able to hand the plane back at the end of the first year – therefore protecting the downside. Boeing agreed.

Because he protected the downside, he was able to tell his fellow directors that at the absolute worst they would lose six months' worth of Virgin record's profits. But if it went well, they would then be able to buy a second and third plane, build another successful business and something they could be really proud of. Because Richard Branson had the downside protected, his board could see the logic of his decision. While they didn't welcome it with open arms, they gave him their blessing and Virgin Airways was born.[16]

All projects involve an element of risk. If you are embarking on a project, then you are taking risks. If you are taking risks, then you are a risk-taker. And, if you are a risk-taker, then you should practice *the art of planning to protect the downside*. The art of planning to protect the downside is what the plan risk management process is about.

Why do we need to plan risk management for our projects?

Projects need to deliver business value. To deliver business value, we need to balance risk and reward. We need to manage risk so that our projects deliver a healthy return on investment.

Creating value while balancing risk and reward is intricate. It isn't something we can do on the fly. We need to plan risk management before we start managing risk if we want to be effective. By doing so, we can more easily identify the threats and opportunities we will encounter when trying to create value.

We need to plan risk management so that we know beforehand how we will balance risk and reward in the course of our projects. And we need to customize each risk management plan so that it is appropriate for each project we embark on.

Each project is unique; hence we need to tailor how we plan to manage risk for each individual project. We customize a risk management plan for each project so that we get the most appropriate balance between risk and reward for each unique project.

The plan risk management process allows you to define how you will conduct risk management for your project. Planning risk management will ensure that the activities that you decide to do are appropriate for your project. And it will ensure that you spend the appropriate amount of time and resources on those activities, executing them to the appropriate level of detail.

Initiate the plan risk management process soon after the project has been conceived. You can then refine the process in the early stages of the project. However you plan your risk management, you should complete your risk planning before you finish the planning stage of your project.

The risk management plan should be completed before you complete the project management plan (PMP) or project initiation document (PID).

Your risk management plan may need to be revisited during the project execution. If something has changed in the project, or if your monitor process identifies areas of your risk management that are not performing well, you many need to update your plan. If the project undergoes a major change in scope or approach, you will likely need to review and revise your risk management plan to ensure it is still effective.

[16]https://www.linkedin.com/pulse/20140224234637-204068115-best-advice-protect-the-downside/

Effective plans are rarely static. Don't be afraid to change your risk management plan if needed. It is in the process of planning that the real value is derived. The plan itself is just the output. Any significant changes to the risk management plan may need to go through the change control process.

The primary benefit of planning is the process of considering all the factors. The actual plan is a secondary benefit. Remember what Eisenhower said:

> *"Plans are worthless. Planning is essential."*
>
> Dwight D. Eisenhower, General in US Army

3.1 Scope of Plan Risk Management

The plan risk management process should reflect your sponsor's appetite for risk. The plan risk management process should reflect that appetite by describing all of the following:

- The risk strategy—the general approach to managing risk in your project
- The methodology that will be used to manage risks for the project
- All the risk management activities your project will engage in
- The risk identification workshops you will conduct and how you will manage them
- What the risk register will contain and how it will be used
- How you will report on risks, to whom, and how often
- How, when, and why risks will be escalated
- How risk management activities will integrate with the rest of the project activities and documents
- How risk related change requests will be handled
- How and how often work performance will be measured, assessed, and reported
- How risk management activities will be funded
- The roles and responsibilities for performing risk management activities

One shortcut I've found to be effective is to ask your executive/sponsor "How should risk be assessed, tracked, and communicated during the project?" They will probably give you high-level motherhood statements, but that will provide a good starting point. Getting their high-level direction at the outset will set you on the right path and will save you time. Asking better questions earlier is the key to finding better solutions.

3.2 Defining Your Sponsor's Risk Appetite

Your sponsor's risk appetite is going to influence the approach that you take to manage the risks of your project. If your sponsor has a low appetite for risk, you'll have to take a more rigorous approach to managing risk. If your sponsor has a high appetite for risk, then you will have some more latitude in your approach to risk management.

"Most innovators are successful to the extent to which they define risks and confine them."

Peter Drucker - American management consultant, educator, and author

To plan risk management appropriately for your project, you need to understand your project sponsor's attitude towards risk. Their attitude will influence how you plan risk management. Speak to your sponsor to determine their risk appetite, risk tolerance, and risk thresholds. Their responses will determine how you approach risk management planning.

Their appetite for risk is unlikely to be spread evenly between the project constraints, and this appetite will be reflected in the risk thresholds that you work out with your sponsor.

How this appetite is split amongst project constraints should be recorded in your risk management plan, as it will inform the actions you take, how you execute those actions, and how often you do them.

We can use ratio scale prioritization methods, such as the *analytic hierarchy process*, to help our sponsor work out how their risk appetite is split amongst the project constraints.

3.2.1 Analytic Hierarchy Process

In project risk management we can use the analytic hierarchy process (AHP) to help us quantify how the sponsor's risk appetite is divided across the project constraints. AHP helps us clarify how our sponsor's risk appetite is spread across costs, schedule, scope, quality, resources and benefits.

The AHP is a structured technique for organizing and analyzing complex decisions. It is based on a combination of mathematics and psychology.

I'm going to explain the method and the benefits. But first, to simplify the process, I'm also going to give you access to a "done for you" tool at www.ictrisk.com. With this tool, all you have to do is input the numbers to reap the benefits of the tool.

AHP was originally developed by Prof. Thomas L. Saaty in the 1970s and has been extensively studied and refined since. It is the most accurate approach for quantifying the weights of criteria. It provides measures of judgement consistency and derives priorities among criteria and alternatives.

In short, AHP is a method to derive ratio scales of multiple criteria from paired comparisons. That may sound complicated, but please allow me to explain. It's simpler than it sounds, and it's a very useful tool.

When we have several different criteria, it is not too difficult to identify a numerical relationship between any two of the criteria. These numerical relationships are referred to as a *paired comparison or pairwise comparisons*.

For example, your sponsor might tell you that schedule is twice as important as cost. And, she might say that cost is 50% more important than scope.

Identifying a paired comparison between two criteria is relatively easy. But, people find it much more difficult to identify the web of relationships between multiple interrelated criteria.

People find it relatively easy to compare two criteria together, but when we compare several pairs of criteria, it becomes much more difficult. When we independently define several pairwise comparisons in a given interdependent data set, we inevitably find mathematical inconsistencies in our paired comparisons.

And that is when AHP comes to the rescue. AHP is a tool that enables you to normalize a set of paired comparisons that are mathematically inconsistent. AHP helps us to make sense of an interconnected set of paired comparisons that inherently don't make sense.

For example, most people would find it very difficult to apply a logically consistent numerical relationship between multiple criteria such as our project constraints: cost, schedule, scope, quality, resources, and benefits.

The AHP is a tool that allows you to input each of the identified paired comparisons (e.g., schedule is 2x as important as cost, cost is 50% more important than scope, etc.).

Any set of pairwise comparisons is likely to have logical inconsistencies that cause wrinkles in your assessment. When you input those pairwise comparisons into the AHP tool, it calculates each of the relationships against one another and shows you the normalized relationship amongst each of the constraints.

AHP is a tool that helps you iron out the wrinkles in your pairwise comparisons to produce a logically consistent set of comparisons.

Table 3 below shows an example of an analytic hierarchy process input table and Table 4 shows an output table (i.e., a normalized table).

Note: in the input table, the amber colored cells reflect the inverse relationship (i.e., if schedule is 2.00 times as important as cost, then cost must be 0.50 times as important as schedule). The amber cells are calculated and do not require user input.

Table 3 - Analytic Hierarchy Process – Input Table

Green Field for Data Input - denotes relative importance of the ROW constraint to the COLUMN constraint						
Amber Cells are Protected Calculated Field						
	Cost	Schedule	Scope	Quality	Benefits	Resources
Cost	1.00	0.50	0.67	1.10	1.25	0.50
Schedule	2.00	1.00	0.80	1.40	1.00	0.75
Scope	1.49	1.25	1.00	1.10	1.25	0.80
Quality	0.91	0.71	0.91	1.00	1.10	1.00
Benefits	0.80	1.00	0.80	0.91	1.00	0.50
Resources	2.00	1.33	1.25	1.00	2.00	1.00
Total	8.20	5.80	5.43	6.51	7.60	4.55

Table 4 - Analytic Hierarchy Process – Normalized Table (Output)

	Cost	Schedule	Scope	Quality	Benefits	Resources	Total	Average
Cost	0.12	0.09	0.12	0.17	0.16	0.11	0.77	12.92%
Schedule	0.24	0.17	0.15	0.22	0.13	0.16	1.08	17.92%
Scope	0.18	0.22	0.18	0.17	0.16	0.18	1.09	18.18%
Quality	0.11	0.12	0.17	0.15	0.14	0.22	0.92	15.33%
Benefits	0.10	0.17	0.15	0.14	0.13	0.11	0.80	13.31%
Resources	0.24	0.23	0.23	0.15	0.26	0.22	1.34	22.34%
Total	1.00	1.00	1.00	1.00	1.00	1.00		100.00%

Say that, in our example, your sponsor has identified the paired comparisons as input in the green cells in the above Input Table.

The Normalized Table then takes the values from the Input Table and converts them into relative percentages by column. It then takes those relative percentages and calculates an average value by row. The values in the *Average* column are the relative values for each row.

In our example, AHP tells us that, given the pairwise input from our sponsor, our sponsor's risk attitude towards *cost* is 12.92% of the total of the six constraints. Our sponsor is clearly far more concerned with *resources,* as the average score for resources is much higher at 22.34%.

We can use this normalized weighting of risk attitude to manage the project's risks according to the average percentage for each project constraint.

We know from our normalized table that the normalized risk appetite for the project constraints is distributed as below.

Cost	12.92%
Schedule	17.92%
Scope	18.18%
Quality	15.33%
Benefits	13.31%
Resources	22.34%

You will notice in the Normalized Table above that the *Schedule* average is 17.92% and *Cost* is 12.92%. This means that the normalized averages show that schedule is only 139% larger than costs, not 200% larger, as our sponsor originally said.

The discrepancy between the pairwise values originally input for *Schedule/Cost* and the normalized averages is due to their relationships with all the other constraints. The relationships between *Cost* and *Schedule* against all the other constraints uncovered a logical inconsistency between the originally input schedule/cost pairwise relationship.

The analytic hierarchy process considers all the inputs and statistically smooths out the inconsistencies to give us a result that are logically consistent.

Sometimes, going through an AHP generates results that are contentious. For example, if you presented the above results to your sponsor, she might say that *Schedule* really is twice as important as *Costs*. In that case, you would simply need to adjust the relative weighting of *Schedule* to the other criteria or the weighting of *Cost* to the other criteria until the *Schedule* is 200% the value of *Cost*.

The free tool that I provide updates in real-time, so you can sit with your sponsor and tweak the paired comparison input values until your sponsor agrees with the results.

The analytic hierarchy process may sound complicated, but you can download the free tool available on our website www.ictrisk.com. Take a few minutes to watch the explainer video and practice with the tool before you use it with your sponsor.

The AHP tool will enable you to determine the *actual* weight that your sponsor places on each of the project constraints. It identifies their pairwise inconsistencies and normalizes them so that they are logically consistent. After you show your sponsor your use of AHP, I would be surprised if your sponsor is not impressed with your prowess!

AHP is specifically mentioned in the PMBOK as one of the quantitative analysis tools. But the PMBOK doesn't go into detail on how to use it. I have found it most valuable in determining a sponsor's risk attitude distribution amongst the project constraints.

Feel free to use the free AHP tool on our website www.ictrisk.com for any other situations in your project when you need to assess the effects of several pairwise comparisons.

3.2.2 Risk Appetite Categorization

Risk appetite is generally described in a categorized range. The risk appetite is often described by categories such as low, limited, moderate, and high. This categorization allows organizations to pre-define risk attitude and risk approaches to match up with each of the risk appetite categories.

An example of how risk appetite categories can match with risk attitude and risk approach can be seen in the diagram below.

Diagram 8 - Risk Appetite Categorization Diagram

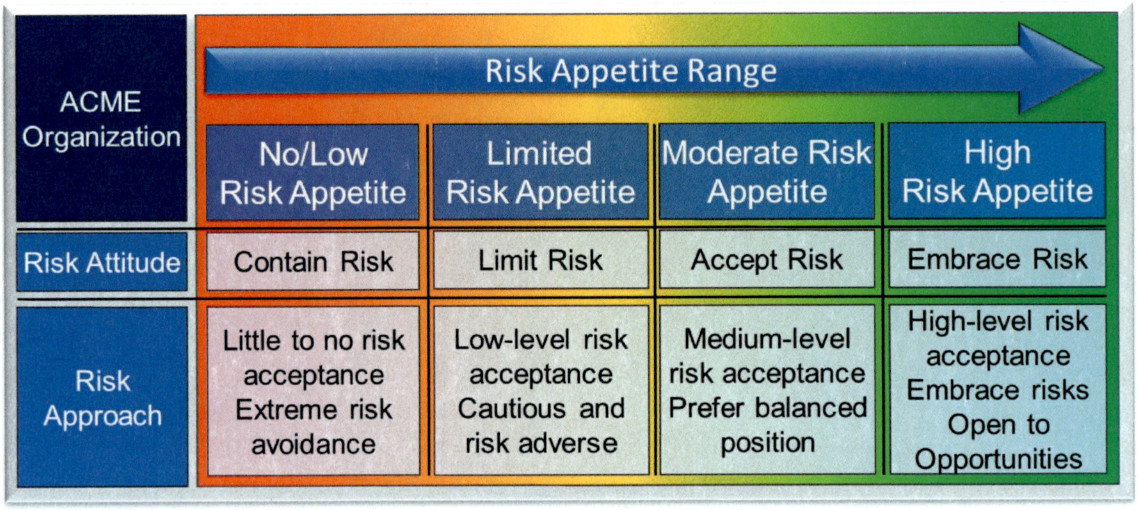

In the example above, I have used four categories to classify risk appetite, but you could use any number of categories you deem most appropriate for your project. This exercise is not definitive. It is simply used as an indicative guide.

3.3 Preparing for the Plan Risk Management Process

When examining your projects to determine an appropriate plan for risk management, we need to consider more than just the risk appetite of our sponsors. We should also consider

1. **Project size.** Is it small enough to justify a simplified risk management plan? Or is it big enough to justify a detailed approach to risk management? When assessing size consider

    - **Budget** – Is the dollar value high? Is it high in relation to other projects? In relationship to your organization? In relation to your business unit?

    - **Duration** – What is the lifespan of your project? How long will it take for your project to deliver? And what is the lifespan of the changes that your project is going to deliver?

    - **Scope** – How broad is the scope of your project? Does it need input from many business units? How much input will it need from business units across the organization?

- **Number of stakeholders** – How big is your project team? How big is the project management team? How big is the project delivery team? How many internal and external stakeholders do you have?

- **Organizational impact** – How big of a change is your project going to have on the organization? How long is that impact going to be felt? How broad reaching is that change's impact going to be? How deeply will people be impacted?

2. **Complexity of your project**. How complicated is your project? How many tasks are involved? How simple or complex are those tasks? What are the interdependencies of those tasks? Can they be performed by individuals or do they require collaboration? How familiar is the technology your project is using? Is it innovative? How complicated are your commercial arrangements? Are there many procurement and contractual considerations? Are there many dependencies in your project?

3. **Project importance.** How important is your project to the organization? To your department? To your business unit? Is it strategically important? How important is your project to your department or your business unit? What impact would project failure have on the organization, your department, or your business unit? What impact would the project's success have on the organization, your department, or your business unit?

4. **Development approach**. Are you using the Waterfall methodology? Can risk processes be followed sequentially and iteratively? Are you using an Agile methodology? If so, is risk addressed at the start of each iteration and during the execution? Or are you using a hybrid methodology which contains elements of both Waterfall and Agile? What impacts will your methodology have on how you manage risks? Whatever methodology you are using, are you covering off both product delivery risks as well as generic project risks?

Make sure that you highlight these considerations to your participants when you conduct meetings to determine how you plan to manage risk in your project.

If you would like to get an idea of your project's risk profile, we provide a *Project Risk Profile Assessment tool* at www.ictrisk.com.

Our *Project Risk Profile Assessment tool* will give you an evaluation of your project's risk profile against industry benchmarks. Our *Project Risk Profile Assessment tool* is available for free for anyone who has purchased this book.

Our *Project Risk Profile Assessment tool* generates a report that you can take away for your consideration or to share. Your free project assessment can be found at www.ictrisk.com.

Meetings are an effective way to further develop your plan risk management process. Involving a broader group of stakeholders in the plan risk management process encourages participants to take ownership of the output. When your participants take ownership, they will be much more likely to adhere to your project's risk management plan throughout the project's lifecycle.

If you involve your stakeholders in your planning and encourage them to take ownership, you may even find that some of them become enthusiastic proponents of your project's risk management plan!

Consider who you select to facilitate your risk management planning meeting. These meetings are often conducted by the PM. But when the PM facilitates the meeting, it diminishes the amount of effort and focus they can contribute to the idea generation component of the meeting.

If you're going to enlist someone else to facilitate the meeting, look for someone who is experienced and skilled at facilitating meetings. Look for someone who has a solid understanding of ICT projects and who has good knowledge of risk management in a project context.

An accomplished facilitator can help participants stay focused on a task, agree on the risk approach, identify and surmount sources of bias, and settle any disagreements that may arise.

3.4 Risk Management Plan

The output of the plan risk management process is a document called the *risk management plan*. The risk management plan defines the strategy and methodology that you will use to manage your project's risks.

The risk management plan should provide all the information relevant to the risk management process. Whether the risk management plan is a stand-alone document or if it is part of another document (e.g., the project management plan or project initiation document) is dependent on how you are running your project.

Regardless of how you intend to document your risk management plan, when you create it, you should keep in mind non-event risks such as variability risk and ambiguity risk. Both concepts are defined below, in line with the PMBOK.

1. Variability risk – uncertainty that surrounds a vital characteristic of a planned event, activity, or decision. For example, the rate of software development may be above or below target, testing may uncover more or less defects than expected, or unseasonable weather may affect change freezes.

   Variability risk can be addressed using quantitative risk analysis tools such as Monte Carlo analysis. In Monte Carlo analysis, the range of variability is reflected in the probability distributions. Then, action can be taken to reduce the spread of the possible outcomes, thereby reducing the risk (see Chapter 6 for more detail).

2. Ambiguity risk – uncertainty exists about what might happen in the future. There will be areas of the project where imperfect knowledge might affect the project's ability to achieve its objectives including elements of the requirement or technical solution, future developments in regulatory frameworks, or inherent systemic complexity in the project.

   Ambiguity risk can be addressed by defining the gap in knowledge and then seeking additional information, either from external sources, simulations, prototyping, or a proof of concept.[17]

[17] PMBOK Page 398, section 11

3.4.1 Risk Management Plan Contents

Typical risk management considerations that you would describe in a risk management plan include

- Risk management strategy
- Risk management methodology
- Risk management activities that will be undertaken in the course of the project
- Risk identification workshop(s) that will be undertaken in the course of the project
- Risk register and its contents
- Risk reporting content and timing
- Risk escalation
- Risk integration with other project processes and artifacts
- Raising risk specific change requests
- Risk related work performance information
- Funding for risk management
- Risk management roles and responsibilities

3.4.2 Risk Management Activities

The risk management plan should list each of the risk management activities that your project will engage in. For each of these activities, the risk management plan should spell out when they will be performed and how often. The document should also state the roles and responsibilities for each of the risk management activities.

I find a RACI matrix is a useful tool to communicate who is responsible, accountable, consulted, and informed (see below for an example of a RACI matrix).

Table 5 - RACI Matrix, Roles and Responsibilities

RACI Matrix										
R	Responsible	Project Manager	Program Manager	Risk Management Consultants	Business Analyst	Test Manager	Deployment Manager	Release Manager	Communications Manager	Organizational Change Manager
A	Accountable									
C	Consulted									
I	Informed									
Risk Identification Workshops		A	C	R	C	C	C	C	C	C
Weekly Risk Reviews		R	A	I	C	I	I	I	I	I
Risk Reassessment Workshops		R	A	I	C	C	C	C	C	C
Risk Reports		A	C	R	C	I	I	I	I	I
Risk Audit		A	C	R	C	I	I	I	I	I
etc.		etc.	etc.	etc.	etc.	etc.	etc.	etc.	etc.	etc.

The risk management plan should outline the *what*, *why*, *who*, *how*, *when*, and *how often* for each of the risk management activities. It should define which of the risk management activities are one-off and which ones are iterative, being performed at intervals during the project lifecycle.

At a minimum, the following risk management activities should be covered by the risk management plan:

- Risk identification workshops – Detail the approach to risk identification activities
- Risk reviews – Detail the approach to risk review activities
- Risk reassessments – Detail the approach to risk reassessment activities
- Risk reporting – Detail the approach to risk reporting
- Risk audits – Detail the approach to risk audit activities

Monitor and control activities - Detail the approach to all monitor and control activities in the risk management plan.

Monitor activities should include a reserve analysis. A reserve analysis assesses both management reserves and contingency reserves.

3.4.2.1 Contingency Reserves

Contingency reserves are either schedule reserves or financial reserves to cover *known* risks. Financial contingency reserves are estimated using expected monetary value (EMV) techniques once qualitative assessments have been completed.

EMV is a simple technique where we multiply the financial impact of a risk against the probability of a risk. For example, consider a project risk of incurring a $100,000 fine for not delivering before a regulated date. If the probability of missing that date is 20%, then the EMV for the risk would be $20,000 ($100,000 x 20%).

As identified individual risks are realized, the cost of the risk should be withdrawn from the contingency reserve account. As individual risks are closed, the EMV for each individual risk can be withdrawn from the contingency reserve account and returned to the project sponsor.

3.4.2.2 Management Reserves

Management reserves are either schedule reserves or financial reserves to cover *unknown* risks.

Gone are the days when organizations launch ICT projects on a hope and a prayer. Nowadays, justification to release funds for an ICT project need to have some rigor behind the estimates.

It is no longer sufficient to stick a finger in the air when determining how much management reserves your project should set aside. Organizations now demand more rigor behind how we determine management reserves.

If you were to simply use heuristic methods to determine management reserves (e.g., "rule of thumb") and you were challenged to explain how you determined them, everyone involved could be exposed to criticism. And, in the event that your project runs into trouble and auditors ask you to explain how you determined the amount of management reserves required, it would be embarrassing to try to defend having simply used a 'rule of thumb.'

If your organization has guidance on calculating management reserves, then use their guidance. If not, in the next sections I will describe a method you can use that has some rigor behind it.

Where you are in the project lifecycle

Projects have a greater amount of uncertainty at the outset. Hence at the beginning, a project requires a larger amount of management reserves to offset the greater uncertainty.

As the project matures, the level of uncertainty drops. As the project moves through the project lifecycle, the level of risk drops. When the level of risk drops, the reserve requirements drop in conjunction.

Generally speaking, management reserve requirements are higher at the earlier stages of the project, and they reduce as the project travels through its lifecycle. But, there are other factors that will influence how much management reserves your project requires.

The Green Book Guidance on Reserves

If your organization doesn't provide guidance on determining management reserves, then the UK HM Treasury *Green Book* provides guidance on determining reserves for ICT projects. It provides a range of total reserves (i.e., contingency reserves + management reserves = total reserves). It provides a range of total reserves and provides a set of assessment criteria to help you determine where in that range your project should lie. You can use this guidance to determine the amount of management reserves for your project.

According to the *Green Book*, ICT project reserves are bounded by the ranges below.

Works durations (i.e., schedule) — 10% to 54%

Capital expenditure (i.e., costs) — 10% to 200%

The *Green Book* provides some guidance to determine where your project sits in the large range between the high of 200% and low value of 10% for cost reserves.

The *Green Book* recommends considering the factors below when determining where your project's cost reserve requirements sit within the high of 200% and low value of 10%. When reviewing the factors, consider how successful have you been to date in

- Identifying possible negative risks in advance? Early risk identification should be used to create contingency reserves which reduces the need for management reserves

- Putting mechanisms in place to minimize the probability of risks materializing with adverse effects? Mechanisms to reduce probability of risks occurring reduce the need for management reserves

- Having the right balance of control in place to mitigate the adverse impacts of risks if they materialize? Mechanisms to reduce impact if risks occur reduce the need for management reserves
- Putting processes in place to monitor risks and to access dependable, up-to-date information? Informed monitor and control processes reduce the need for management reserves because they allow you to react quickly to risks and to offset their effect
- Having decision making processes supported by a framework of risk analysis and evaluation? When decisions are made within a framework that informs the decision maker of the risk implications, there is less of a need for management reserves
- Consulting early and in a detailed manner with the majority of your stakeholders? Early and extensive consultation with stakeholders flushes out requirements in more detail and reduces the likelihood of scope surprises, thus reducing the need for management reserves
- Avoiding making irrevocable decisions and having a complete assessment of costs, including the potential to delay decisions? Avoiding irrevocable decisions increases the options available and reduces the need for management reserves
- Conducting pilot or proof-of-concept studies? Getting more information about the risks faced by a project through pilots or proofs of concept allows projects to take action to mitigate risk or increase the benefits, thus reducing the need for management reserves
- Introduce design flexibility? When future demand and relative prices are uncertain, a flexible design reduces the need for management reserves
- Adding a precautionary principle? Can precautionary action be taken to mitigate risk? Introducing precaution as a principle reduces the amount of management reserves required
- Assessing procurement contractual risk? Risk that can be transferred to other parties through contracts reduce the need for management reserves, e.g., insurance
- Use proven, rather than leading edge, technology? Risk is lower when using technology that the organization and/or project team have used before, thus reducing the need for management reserves
- Able to reinstate or develop different options? Following the risk analysis, developing alternatives that are either less inherently risky or those that deal with the risks more efficiently, reduce the need for management reserves

If that seems difficult to assess, I have developed a reserve calculator based on the above that you can use to help you determine how much management reserves your project should hold as contingency.

Even if your sponsor isn't going to set aside any more management reserves, it is still worth running your project through an assessment. The output of the assessment will give you some evidence to defend yourself if the reserves you've been given prove to be inadequate.

If nothing else, the reserve calculator gives you a mechanism to show your sponsor how much management reserves industry best practice recommends. That way, if your sponsor provides management reserves that prove inadequate, you have something to fall back on. You can find our free *management reserves assessment tool* on our website at www.ictrisk.com.

As your project progresses and *unknown* risks are realized, the amount of money you spend on dealing with the ensuing issues should be withdrawn from the management reserves account. Any remaining management reserve funds stay in the management reserve account until the project is closed.

3.4.3 Risk Identification Workshop(s)

Risk identification workshops focus on the identification and evaluation of new risks. These workshops are especially important at the early stages of a project. But risk identification and evaluation are required throughout the project to capture and assess any emergent risks.

These workshops can be supplemented with interviews (often for executives and specialists) and with questionnaires (especially for remote stakeholders in awkward time zones).

To define how you will manage risk workshops, this section of the risk management plan should outline

- Strategies for managing threats – the high-level strategy for how the project will manage individual risks that have negative impacts
- Strategies for managing opportunities – the high-level strategy for how the project will manage individual risks that have positive impacts
- Strategies for managing overall project risks – the high-level strategy for how the project will manage the risk of the project as a whole

3.4.4 Risk Register

The risk register captures details of individual risks identified. The output of each of the stages in the risk management process are recorded in the risk register.

This section of the risk management plan should define what information will be included in the risk register. The risk management plan should define each field the risk register should contain, including

- Unique identifiers – a unique identifier for each risk. Don't reuse identifiers if risks have been closed or otherwise removed. Ensure each risk has a unique identifier regardless of the risk status
- Short risk title – this should be something long enough to be unique and give meaning, but short enough so that you can easily reference it in a conversation
- Risk description – this should be a complete description of the risk. It should follow risk metalanguage (cause + event + effect). The level of detail of the description should relate to

the importance of the risk. A risk with a *low* risk score may be given a concise description while you would likely go into much more detail for a risk with a *high* risk score

- Risk category – organizations often have set categories that they use to group risks. In some organizations, these come directly from a risk breakdown structure (see glossary)
- Risk status – you might have multiple statuses for risks. At a minimum, you should have *open* and *closed*. Some organizations use the terms *urgent* and *watch* instead of *open* to differentiate the risks that are important enough to be actively managed on the urgent list and those that are less important, but still deserve tracking on the watch list
- Risk trigger - events or conditions that indicate a risk is about to occur; especially important if you use contingency plans in your risk response plan
- Risk owners – named stakeholders who are responsible for individual risks
- Risk actionee – a person who will be executing action as part of the risk response plan
- Risk response plans – active or passive activities that will be undertaken to treat individual risks
- WBS references – unique identifiers to related tasks in the work breakdown structure (WBS)
- Timing information including when the risk was identified, when it might occur, when it would become irrelevant, and when the deadline is for taking action.[18]
- The risk register should record impact and probability – it should record *inherent* risk impact and probability, then list the risk response plan, and then list the *residual* risk impact and probability
- Risk score for both inherent and residual risks
- The risk register should also have the capacity to record any secondary risks and their impact, probability, and risk score

If useful for your project, you can also include in your risk register the following information:

- Target risk values – the target scores for impact, probability, and risk scores; these can be used to evaluate the effectiveness of risk response plans
- Urgency of the risk – how soon a risk response needs to be applied to treat the risk
- Root cause – the underlying reason that the risk exists
- Warning signs / symptoms – indicators that the risk is about to occur or has already started to occur
- Risk proximity – the amount of time between the present and when the risk could occur

[18] PMBOK Section 11.2.3.1 Risk Register page 417

- Notes for risks that need further analysis
- Contingency plans – risks whose response plan includes a predetermined reactive set of actions should the trigger be enacted
- Fallback plans that can be used should the risks occur, and the primary risk response plans prove ineffective
- Monitoring and controlling actions for assessing the effectiveness of risk response plans

The risk register is fundamental to the risk management process. The risk register is used from the initial risk identification through to project closure. Think of the risk register as the backbone of the risk management process with all activities but the plan risk management hanging off the risk register.

Ensure that your risk management plan includes all the risk register information relevant for you project.

You can download a template of a fully formatted *risk register* for free at www.ictrisk.com.

3.4.5 Risk Reporting

Risk reporting is about communicating your project's risks to your stakeholders. They will want to know about the overall project risk, AND they will want to know about the individual risks. In the risk management plan, you outline the strategy you will take to communicate your project's risk information.

When describing risk reporting, you need to consider

- What risk information are you going to share?
- What level of detail will your stakeholders want to see? Will they be happy with simple traffic light reporting (e.g., how many risks are in each level of severity—red, amber, green)? Or will they want to know the detail? If so, how much detail?
- How will you communicate risks to your stakeholders? What medium will you use to communicate? Will it be a templated report that you email? Will you maintain a project dashboard that they can look at whenever they want?
- Who will you send which pieces of information? Will you have customized reporting for different stakeholder groups that you send different levels of information?
- What level of risk information is appropriate for each of your stakeholder groups? Will you send the same risk information to your external stakeholders that you send to your project board?
- How frequently will you report on your risks? Will you report weekly? Monthly? Will you report on different time intervals for different stakeholder groups? Will you regularly update a dashboard?

3.4.6 Risk Escalation

In the risk management plan, you should define how you will escalate risks. This is an important piece of planning that often gets overlooked.

Defining your strategy for risk escalation forces you to think through parts of the process that are often not considered until later in the project. Working out how your risk escalation process will work before you need it will prevent confusion and delay later in the project when escalation is required.

You need to define *when* you will escalate risks. Specifically, you need to define the criteria and tolerances that will trigger risk escalation. Obviously, this can't be decided in isolation. The sponsor, program manager, or board that you will be escalating to will have an interest in the escalation triggers and tolerances. You should consult them when defining risk escalation.

When defining your escalation criteria, you should be explicit in stating what would cause you to escalate a risk. There are many criteria that could trigger an escalation. Consider if you would escalate on the basis of

- An individual risk score exceeding a specific metric
- The number of risks in a particular category exceeding a specific metric
- The overall project risk exceeding a specific metric
- The results of your reserve analysis exceeding a specific metric
- A subjective assessment of your ability to manage the risks dropping below a pre-defined capability
- A subjective assessment of the effectiveness of your risk management plan dropping below a pre-defined level of effectiveness
- A subjective assessment of the effectiveness of your risk response plans dropping below a pre-defined level of effectiveness

You need to define who you will escalate risks to. Will all escalations follow the same escalation path? Or, will you have different escalation paths dependent on the types or severity of risks? Or, will you have different escalation paths dependent on the trigger criteria?

And you need to define how you will escalate risks. You may want to define how you will escalate risks dependent on the risk score, the severity of the impact, or the proximity of the risk. You might create several levels of criticality that define how you escalate risks.

Can a risk escalation wait until you can write up a risk escalation report? Or, does your risk trigger justify phoning your sponsor at home outside of office hours? Those are the types of questions your risk management plan should answer.

Defining your strategy for escalating risks before you need to escalate a risk will ensure that everyone involved is clear on what will be escalated, when it will be escalated, and how it is to be escalated.

That way, if an escalation is triggered, everyone will know what to do and everyone's expectations will be set.

3.4.7 Risk Integration

Risk management needs to be integrated into the other project processes. The risk management process provides limited value if it is executed in isolation. The output of risk management processes needs to be incorporated into other project activities and documentation.

In the risk management plan, outline your strategy for integrating the risk management process into the other project processes and documentation. You need to define what elements of the risk management process will integrate, how they will integrate, when they will integrate, and who is responsible for the integration.

You need to consider how risk management activities, like risk response plans, integrate with your other project activities and documents. How will contingency plans integrate? How will fallback plans integrate? How will risk audits integrate? How will reserve analysis integrate?

Consider each of the activities in your risk management plan and identify which project activities and documents that they will integrate with and how they will integrate.

Risk management activities need to be considered when revising the project plan. And you need to define what other project documents need to be updated, how they will be updated, when they will be updated, and who will update them.

Take the time early in your project to determine your strategy for integrating risk management into your other project activities and define it in your risk management plan. By doing so, you'll avoid confusion, misunderstandings, delays, and rework.

3.4.8 Raising Change Requests

Some risk management activities will exceed your change tolerances. You will have to raise a change request for these activities. In the risk management plan, you should define your strategy for how you will handle the raising of risk related change requests. All stakeholders should be clear on how the process works.

For example, if your risk response planning comes up with a response that is outside of your change tolerances and requires a change request, how will you handle it? In the risk management plan, you should define when, how, and by whom change requests will be raised.

You may decide that you will simply piggy-back off your standard change management process. Or, you may decide you need a custom process for risk related change. Whatever you decide, you should define the process in your risk management plan to avoid confusion, misunderstanding, and rework.

3.4.9 Work Performance Information

Despite our best endeavors, our risk management plans will never be perfect. We will need to plan to monitor and control how effective our risk management plan is. We need to consistently monitor how effective our risk management efforts are, and we will need to adjust our plans and methods accordingly.

Your risk management plan should define the strategy you will take to monitor and control risks. It should define the *what*, *who*, *how*, *when*, *why*, and *how often* of your monitor and control activities.

Your risk management plan should define

- What actions are to be taken to monitor and control your project risks
- How you will perform monitor and control activities
- When monitor and control activities are to take place
- How often you will perform monitor and control activities (if they are iterative)
- Who will perform the monitor and control activities? Who is responsible? Who is accountable? Who will be alerted?
- What metrics you will use to monitor risks, risk responses, contingency reserves, management reserves, and overall risk management process effectiveness

3.4.10 Funding for Risk Management

The risk management plan should provide information on how risk management activities will be funded. It should address questions such as

- How are the funds for risk management activities managed?
- What is the process for requesting and accessing the funds?
- How and when will the funds be released?
- How will the expenditure be tracked and reported?
- What are the protocols for the application of contingency reserves and management reserves?

Even if you have covered each of those questions elsewhere, you should address them in the risk management plan as well. If they have been covered elsewhere, you could simply provide a reference to the source document so that a reader of the risk management plan knows where to find risk management funding information.

3.4.11 Roles and Responsibilities

The risk management plan should outline the roles and responsibilities for each of the risk management activities. For projects that are relatively small, straightforward or that the team has done many times before, the PM and project team will often have the capability and capacity to manage the risk management activities in-house.

But, if projects are large, complex, or introduce unfamiliar technology, the project team may not have either the capability or the capacity required to manage the risks. Important projects often justify bringing in project risk specialists who have expert risk analysis capabilities. And bringing in project risks specialists ensures the project has the capacity to dedicate enough time and effort to managing the project risks well.

In important projects, it can often be appropriate to bring in third-party expertise to help manage project risks. On important projects project managers are often too busy managing the projects and the inevitable issues to give risk management the time and attention required. PMs are often too busy herding cats and fighting fires to give risk management the time and dedication it requires.

3.4.11.1 Benefits of Using Third-Party Project Risk Management Expertise

There are four arguments for considering external expertise to help with project risk management. You can address these concerns in the risk management plan so that your stakeholders are clear on the approach you are taking and why.

The first reason to consider engaging third-party risk management expertise is the limited capacity of the project team. There are going to be points in any project when the project team is spread so thin that they simply can't spend an adequate amount of time on risk management.

Most projects go through a phase (or several phases) where important issues flare-up and make huge demands on the project team's time and attention. If we knew when those times would be, we could plan to accommodate them. But we don't know. So, engaging help with risk management early in the project will ensure you have dedicated support when you need it.

If your project is important, you can't afford to neglect risk management during these phases. If you do, you may put the project objectives in peril.

It is in those hectic phases of a project that risk management is often neglected. A PM can be kind of like a homeowner with a kitchen fire.

When there is a fire in the kitchen, a homeowner isn't going to check to see if the smoke detectors in the bedroom are working. So too, PMs are not inclined to check on risks that *might happen*, if there is one that *has happened.* If there is a burning issue, it gets the priority and smoldering risks get ignored, leaving the project vulnerable.

A risk that *has occurred* has become an issue. And issues have consequences that need to be dealt with immediately. That immediacy crowds out the needs of unrealized risks. The danger is that the smoldering unrealized risks go unchecked, until one of them bursts into flames unseen.

The second reason to consider engaging third-party risk management expertise is the need for the PM to stay positive and focused on delivery. Managing a project is all about delivering an outcome for the customer. The PM should be aware of risks, but their primary focus should be positive and should be applied to project delivery.

There is a danger that if the PM focuses too much attention on threats, the PM will adopt a negative mindset. A negative mindset can introduce pessimism that adversely affects project progress.

Applying too much focus on negative risks can change your mindset through the reticular activating system (RAS).

RAS is a network of neurons that filters out the unimportant things that you see and hear. It does this so that the important things can get through to your brain. Your RAS takes what you focus on and creates a filter for it. It then sifts through the data that your eyes see, and your ears hear. The RAS then presents to your consciousness only the portions that are important to you (i.e., the things that you have been focusing on).

The RAS works subtly, without us ever being conscious of it. Because of the way that the RAS works, relying on your PM for all of your threat management can cause them to view project delivery with an air of negativity and pessimism. Please see Appendix C for more on how the RAS works.

The third reason to consider engaging third-party risk management expertise is capability. Third-party risk management specialists have typically focused years of their time and attention on developing an expert risk management capability.

PMs need to have a broad range of skills to effectively manage projects. They need to be good at project integration, scope management, schedule management, cost management, quality management, communications management, risk management, procurement management, stakeholder management, and much more.

But, when you have to be "good" at so many different things, you never have the time to become "great" at any one of them. A PMs time is spread so thin over so many areas that they just don't have the time to specialize in any one thing. Nor should they. A PM should be a generalist that has a good enough knowledge of many different areas so that they can coordinate and manage different specialists to successfully deliver a project.

And, rarely are PMs experts in all areas of risk management. After reading this book, you will be able to manage project risks well. But, even if you absorb and apply all the information in this book, you are still going to have gaps in your risk management knowledge. For example, you are unlikely to have the software simulation tools and the knowledge to use those tools to apply the more advanced quantitative risk management techniques.

It is rare to find a PM who has invested the funds on purchasing Monte Carlo simulation software and the time in learning how to run and analyze Monte Carlo simulations. For most PMs, there are simply too many other aspects of project management to justify focusing on something that they might only use a few times per year. Monte Carlo simulations are typically left to the specialists, or simply not performed at all.

On a project that is large, complex, or introduces a new technology, it is common to have specialists brought on board to help with tasks that a PM would typically do in a small project. Third-party specialists provide a deeper level of specialist capability than could reasonably be expected of even the best educated and most seasoned project manager.

On an important project, you might have a specialist scheduler brought on board to help with the schedule. You might also have a specialist communications manager to help with communications. And, you might have a specialist change manager and a specialist benefits manager. All these experts specialize in a specific project subject and contribute with deeper knowledge and experience in their area. They provide expertise that a PM simply doesn't have the time to acquire. The same logic applies to engaging a project risk specialist.

The fourth reason is that external third parties bring a broad and fresh perspective that has not been prejudiced by widespread assumptions.

When we are engrossed in a project it is common to get so immersed in the details that we lose perspective of the bigger picture. We get so caught up in the detail that it can appear out of proportion with the project as a whole. As the old idiom says, "Can't see the forest for the trees." An external third party brings a view that is fresh and unsullied. An external third party brings a perspective that is more proportionate.

When working on a project for an extended period of time, we can become insensitive to widespread assumptions. Whether assumptions have been explicitly defined or if they are subtle and presumed, widely held assumptions can conceal risks of their own. External third parties bring perspective that is more open, and they are more inclined to challenge defined assumptions and uncover presumed assumptions.

If your project is large, complex, or introduces a new technology, consider bringing in a third-party project risk expert.

If you do have a risk manager within your organization, consider seeking their assistance. Before you do, it is worth asking them how much project risk management experience they have. This is because, although the principles are the same, the execution of project risk management is different from enterprise risk management or financial risk management.

If your risk manager doesn't have project risk management experience, consider engaging the services of a third party who specializes in project risk management. On an important project, you want to be seen doing all that you reasonably can to manage project risk and ensure that you deliver your project objectives.

Whether you decide to engage a third-party specialist or not, you should document your approach and rational in the risk management plan.

The risk management plan should outline all the risk management activities so that your relevant stakeholders know what will be required to manage your project's risks.

3.5 Chapter Summary - Plan Risk Management Summary

The plan risk management process should outline all the activities that you will engage in to manage risks for your project. The project management plan is the natural output of the process. Every project is unique. The project management plan should outline the *what*, *how*, *why*, *who*, *when*, and *how often* of risk management specific for your unique project.

The risk management plan should reflect your sponsor's risk appetite. It includes details on

- The risk strategy and the reason for adopting the strategy
- The risk management methodology that your project will use
- The risk management activities that your project will engage in
- The risk workshops that you will conduct
- The risks register and its format
- Risk reporting
- Risk escalation
- Risk integration
- Risk related change requests
- Risk related work performance information
- Funding for risk management
- Risk roles and responsibilities

The risk management plan should include details on how risks will be identified and systematically assessed, iteratively, and collaboratively throughout the project lifecycle.

4 Risk Identification

Risk identification is the second link in our risk management process chain of steps.

Diagram 9 – Risk Identification

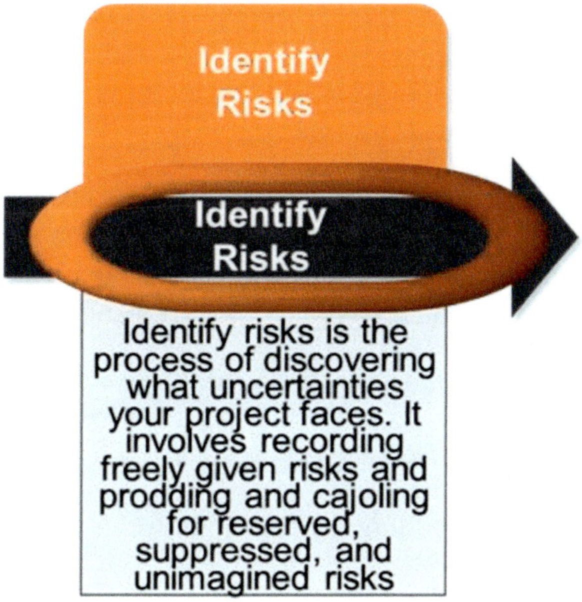

The identify risk step concentrates on identifying individual project risks. Risk identification is crucial to the manage risk process. Any errors or omissions in the risk identification process cascade through the entire risk management process.

If the risk register is the backbone of the risk management process, then risk identification is the heart of the process. Risk identification pumps individual risks as an input into the risk management process just as a heart pumps life giving blood through the body.

But, despite being vital to the risk management process, risk identification is not without its challenges. Fortunately, the risk management industry has developed tools and techniques that we can use to overcome those challenges and effectively identify individual project risks.

4.1 Challenges to Risk Identification

> *"It's not because things are difficult that we dare not venture. It's because we dare not venture that they are difficult."*
>
> Seneca – Stoic philosopher

The complexity of risk identification tends to be underestimated. The obvious nature of many project risks can obfuscate the difficulty in unearthing the few risks that are truly dangerous and difficult to identify.

The ease of identifying the many, easily identified risks obscures the difficulty in identifying the few, opaque, and dangerous risks.

Projects often have some glaringly obvious risks. Nobody would go to the effort or expense of running a risk workshop to gather these low-hanging fruits. These are the risks that are obvious to almost everyone involved. No, we conduct risk workshops to find the opaque, dangerous risks that are difficult to identify.

Project risk identification is kind of like sailing. When sailing a boat, it's not the islands that the sailor is concerned about. Islands are easy to spot and simple to avoid. It is the rocky underwater outcrops, shoals lurking just under the waterline that pose the real threat to the sailboat. That is where the real danger lurks, and those are the risks that the sailor carefully watches for.

So too with individual project risks. It is not the obvious risks that pose the real threat. They are like the islands, easy to spot and simple to avoid. Rather, it is the subtle and opaque risks that pose the real peril. They are like shoals lurking below consciousness. These are the risks that tend to catch out the unwary. These are the risks that pose the biggest challenge and the most danger to projects.

Identifying the easy and obvious risks is useful as a warmup exercise for your participants during the risk identification process. But, the real challenge will be identifying the subtler risks, the obscure risks that could blindside you if you're unwary.

Awareness of these subtler risks lurks in the recesses of the minds of your workshop participants. Your task is to expedite the journey these risks must make from the minds of the individuals to your project risk register. That journey begins with teasing these risks to the surface of consciousness. Then, we need to encourage our participants to share them with the group. And then, we need to define and refine them with the group, continuously overcoming obstacles along the way.

Without risk management tools and techniques, it would be very difficult to identify all but the obvious risks. Our brains developed in a simple environment, where risks were much more immediate and visible than the environment we live in today. The majority of human existence has been in a world far simpler than the world of today. We're simply not naturally endowed with the mental capability to identify complicated risks impromptu.

The development of the human brain has not kept up with the technological and business advances and of the last few centuries. Consequently, humans today are inherently ill-prepared for identifying and assessing risk in the complicated, swift-moving and interconnected world of today. That is why we need tools and techniques to guide our thinking about risk management.

Human nature is going to be one of the greatest challenges to your project risk identification. Most of your workshop participants will need to work their brains hard to nominate any risk that is not immediate and visible. Using a selection of tools and techniques to coax, prompt and cajole your participants will be your best approach to overcoming the challenges that human nature poses to your workshop.

The risk identification tools and techniques that will help you to succeed in your workshop are described on the following pages. These tools and techniques have been extensively tested and proven by the risk management industry. These tools and techniques will help you overcome the risk identification challenges that your project faces.

There are four main areas of human nature that create the challenges that PMs face when conducting a risk workshop. These are

1. Reserved or cowed participants,
2. Mindset bias,
3. First mover bias, and
4. Hijackers.

I address each one of these challenges in more detail below.

4.1.1 Reserved or Cowed Participants

"The flower that smells the sweetest is shy and lowly."

William Wordsworth - Poet

So too with your participants. Your confident, outgoing, extroverted participants are going to help you quickly identify some risks. But, you will likely find that some of your most important risks are identified by your participants who are shy, contemplative, or simply less inclined to speak out in a group setting.

Getting stakeholders to even attend a risk workshop is a major challenge. Fortunately, most organizations are aware of this and at least pay lip service to encouraging participation. Some techniques to encourage participation are more successful than others. A broad dictate that all stakeholders must participate isn't very effective.

I once attended a risk workshop where a broad dictate approach was employed. When the number of risks being volunteered dried up, the facilitator pointed a finger at an unsuspecting stakeholder and demanded, "You! Give me a risk!" Such an approach is unlikely to yield any worthwhile results.

Participants need to be coaxed, prompted, and cajoled into thinking through the project unknowns. They need to be led on a journey, not shoved down a path. If we are going to get any worthwhile output from our workshops, we need to move beyond dictating that all must participate. We must engage our participants, and we need to entice them to contribute.

To entice our stakeholders to contribute, let's first look at common reasons why people keep their ideas to themselves in group settings. Then, we will go through methods we can use to persuade them to consider uncertainty more deeply. And, we explore methods to stimulate and encourage them to share their ideas.

Political Considerations

First up are the political considerations that can discourage stakeholders from participating. Risk workshop participants often span the organization chart — from entry-level employees to senior management. Participants may feel cowed into silence by any of the mountains on the political landscape. It can take a brave junior to speak freely in front of the executives. Most junior employees are not so inclined.

Yet, that is precisely what you want in your workshop. Workshop brainstorming should be unconstrained "blue sky" thinking where all kinds of ideas are suggested for consideration. We need to create workshops that are democratic, workshops where all opinions are considered, regardless of the source.

To miss out on the contribution of junior members would lessen the quality of your risk identification. Senior members may have a broader, more interconnected perspective, but more junior participants are often more aware of the detail. Both perspectives are required to unearth a comprehensive range of project risks.

If we are going to reap the full benefits from the workshop, we need to encourage our participants to contribute any and all ideas, without any self-censorship.

Just as comments that are clever and insightful are welcome, irrelevant comments may also be helpful. Irrelevant comments can prompt lateral thinking. You never know when an irrelevant comment is going to trigger a tangential thought process that identifies new relevant risks. In a risk identification workshop, there is no such thing as a dumb idea. And there is no such thing as a dumb question.

Unfortunately, too few participants will readily share uncertain ideas for fear of embarrassing themselves in front of colleagues, their boss, or their boss's boss.

Many people lower on the organizational chart are inclined to keep their thoughts to themselves. Or they will only bring up the "safe" and obvious risks. Or they will stick to risks that they know management already have on their mental radar.

This phenomenon is not restricted to vertical relationships in the org chart. Two people at the same level who are jockeying for position may reserve their comments for fear of saying something stupid in front of managers senior to them. They may stifle their participation for fear of giving their competitor an advantage.

Excessive Thoughtfulness

Another reason for people not contributing during workshops is the nature of their thought process. Some people prefer to mull over ideas in their heads. They prefer to perfect their ideas before voicing them.

Such participants are less likely to speak up. I don't consider this a negative characteristic, but it is something that you will need to manage if you're going to get contributions from everyone in your risk workshop.

When it comes to risk identification, it is far better for people to contribute ideas that are only partially developed than to hold them back until ideas are fully formed. In the fast pace of a risk workshop, so many ideas are considered that any ideas set aside to be perfected are often lost before they can be reconsidered. We want participants to share their ideas, even if they are flawed. As Confucius said:

> *"Better a diamond with a flaw than a pebble without."*
>
> Confucius

Shy or Reserved Participants

And finally, there are those people who are inherently shy or reserved. If you didn't know this about one of your participants before inviting them, you will likely soon find out. If you manage to draw these wallflowers out and engage them in the workshop, you will get a more complete view of the risks your project faces.

The tools and techniques to address reserved or cowed participants are dealt with later in this chapter.

4.1.2 Mindset Bias

A person's inclination to focus on a particular subject at the exclusion of other subjects is known as *mindset bias*. It is another challenge you need to be aware of in your risk workshop.

> *"The eye sees only what the mind is prepared to comprehend."*
>
> Robertson Davies, Tempest-Tost Author

A powerful cause of this preoccupation is a bias towards the area that a participant is responsible for and/or has expertise in. People often concentrate their mental efforts on the subject matter they are responsible for, influence, or have an interest in. For example, someone from finance would naturally bring a mindset that primarily focuses on costs.

Participants have the ability to add value to areas outside of their mindset bias. But you will have to entice and encourage them to also apply their knowledge and mental focus to the other aspects of the project.

> *"We all see only that which we are trained to see."*
>
> *Robert Anton Wilson,* Masks of the Illuminati

The second cause of mindset bias is the subject that your participants have been focusing on before the workshop. When a person has been absorbed in a particular subject for a prolonged period, it can take them a while to swing their mental focus away from that subject and onto a new subject.

People tend to continue thinking about a subject that they have been focusing on long after they have stopped focusing on it. The inclination of the mind to persist with a subject long after they have stopped focusing on it poses a challenge for the identification of project risks.

Take the example of a finance stakeholder who arrives at the workshop mentally preoccupied with the end-of-month reporting that she has just been working on. Her mindset may remain focused on reporting for much of your workshop.

She may only contribute risks that relate to finance or reporting. But if you can broaden her focus away from her mindset bias and entice her to apply her own particular way of thinking to other aspects of the project, she may have unique insights to offer into other areas of uncertainty.

And the mindset bias is even stronger when there is a large distinction between the subjects people are asked to switch to.

For example, a person's mindset when working in the abstract, such as the creative design of an advertisement, is quite different from when they are working on something concrete and structured, such as a mathematical calculation. There is a switch cost when we move from one subject to a distinctly dissimilar subject.

4.1.2.1 Switch cost

Cognitive shifting is the mental process of *consciously* redirecting one's attention from one fixation to another.[19] Cognitive shifting is required often when participating in a risk workshop. But, cognitive shifting comes at a cost. That cost is called the *switch cost.*

Performance on tasks is disrupted and diminished when a switch from one task to another is required. This reduction in performance is the switch cost. This disruption is characterized by a slower performance and a decrease in accuracy on the new task.

The difference in accuracy and performance between a task repeat (A-A) and a task switch (A-B) is the switch cost. The switch cost remains even when there is ample warning of an upcoming switch, thus it is thought to reflect the functioning of numerous executive control processes ranging from attention shifting, goal retrieval, task set reconfiguration processes, and inhibition of prior task set.[20]

The longer we have been directing our minds down a particular mental alley, the longer it can take to swing our focus down a different mental alley. The longer we have been focusing on one subject before being asked to consider another subject, the greater the switch cost.

The speed at which people can change mental focus varies from person to person, but we are all subject to some delay in exercising this mental agility. The speed of your participants' cognitive shifting will vary from person to person, but it is something you will need to consider when planning your workshop.

Workshop participants who have been working on a particular focused task before the workshop will have some degree of mindset bias. A manager who has spent four hours before the workshop concentrating on resource issues will likely come into the workshop with a mindset bias toward resource management.

That manager is less likely to come up with risks relating to other areas of uncertainty than if she had come into the workshop before starting her other work. There will likely be a delay in optimal cognitive focus while she mentally refocuses. The concept is depicted in Diagram 10, below.

[19] https://en.wikipedia.org/wiki/Cognitive_shifting
[20] https://en.wikipedia.org/wiki/Task_switching_(psychology)#Switch_cost

Diagram 10 - Mindset Bias Diagram

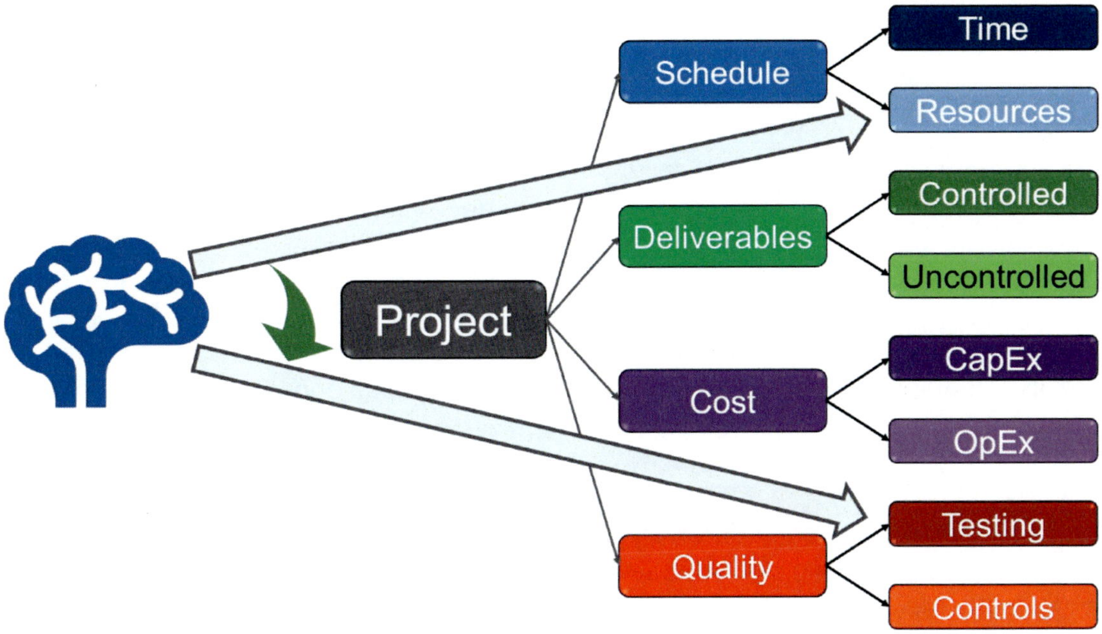

> *"I think that no matter how smart, people usually see what they're already looking for, that's all."*
>
> Veronica Roth, Allegiant

The facilitator of the risk workshop should organize the agenda to mitigate the effect of mindset bias and to speed cognitive shifting. The facilitator must be cognizant of the switch cost as they lead the workshop participants into probing every category where risks may lurk.

4.1.3 First Mover Bias

The tendency of a group's thinking to skew towards the first topics suggested is known as *first mover bias*. Over the years, I have observed this phenomenon in risk workshops many times. As soon as the floor is open to ideas, whichever subject is articulated first significantly influences the nature of the output of the rest of the workshop.

When a risk identification session is initiated, participants' minds are the most open. They are thinking broadly about uncertainty. In this open state of thinking, people's thought processes are susceptible to suggestion. The first subject suggested acts like an anchor for the rest of the workshop. Conversation may drift out around the anchor, but it tends to be constrained by the anchor.

If, for example, someone initially proposes a risk that relates to costs, you will find the workshop produces significantly more risks related to costs. If the first risk suggested relates to a quality, you will find significantly more risks proposed related to quality, and so on.

The skewing of the collective thought process by first mover bias is normal. Fortunately, as the facilitator, you can offset first mover bias by using the tools and techniques in this chapter. And, you can make a major contribution to resetting the collective mindset after each idea is contributed by providing context and outlining categories. These tools are described in more detail throughout this chapter.

4.1.4 Hijackers

There are numerous ways a workshop can be hijacked. Hijacking of a workshop is the monopolization of the workshop for purposes other than those intended. Hijacking can stem from several causes. Regardless of the cause, it is the facilitator's challenge to minimize any monopolization of the workshop, to keep to the agenda, and to achieve the objectives of the workshop.

As with mindset bias, participants often bring to a workshop unrelated issues that have been occupying their thoughts. Two participants who may be having an ongoing debate, about an unrelated subject, may digress to that subject every time your workshop comes anywhere near the disputed topic.

As a facilitator, you may already be aware of participants who have been guilty of hijacking in the past. Identify the likely suspects at the outset and be ready to quickly redirect them back to the task at hand. But the "likely suspects" are not the only ones you need to be wary of.

You should be prepared to step in to arrest a hijacker at any time. Occasionally, a person who has never spoken up before will veer off topic and start a tangential conversation, hijacking your workshop. Later in this chapter, I outline tools and techniques you can use to defend your workshop against this time-wasting behavior.

4.2 Preparation

> *"By failing to prepare, you are preparing to fail."*
>
> Benjamin Franklin - Polymath

Preparation is the key to conducting successful risk workshops. Preparation is key to ensuring our workshops achieve our objectives and don't waste resources.

Workshops are expensive. Regardless of whether your workshop incurs external costs for facilities or resources, the opportunity cost of taking so many people away from other productive activities is significant. You need to ensure that the value that your workshop provides far exceeds the costs that you incur.

Preparation is the key to ensuring that your workshop delivers value. Preparation ensures that you are not seen to waste the organization's time and money. Take a logical and structured approach to preparing your workshop to achieve your desired outcomes. A logical and structured approach helps avoid wasting money. And it can prevent a loss of face.

The following is a step-by-step preparation process that will greatly enhance your chances of successfully achieving your workshop objectives.

Risk workshops give you significant exposure to a broad range of stakeholders. And, by using the preparation approach described below, you give yourself a chance to shine in front of stakeholders of influence. If you utilize the tools in this book, you will be remembered as a well-prepared, logically organized, and effective PM.

4.2.1 Preparation Process

> *"Give me six hours to chop down a tree and I will spend the first four sharpening the axe."*
>
> Abraham Lincoln – American President

As with chopping down a tree, thorough preparation is the key to facilitating an outstanding workshop. Preparing for the workshop can be broken down into seven steps:

1. Select participants
2. Send invitations
3. Structure the agenda
4. Send pre-reading
5. Create the presentation
6. Gather material
7. Prepare the facilities

These steps are explained in more detail later in this chapter.

4.2.1.1 Participant Selection

Your first step in preparing for your workshop will be to select the participants. While it may seem appealing to have all and sundry participate in your workshop, there is an optimal range. If you go beyond this optimal range, the exercise becomes expensive, time-consuming, and ineffective.

Try to include representatives for each project phase and functions. One participant to represent each of the key deliverables should be included. Review the scope of your project to determine how many unique deliverables you have and ensure each unique deliverable is represented. If the number of participants starts to add up, remember it is okay to have one representative covering several deliverables.

Where different skill sets are called upon to produce the deliverable, try to have at least one representative for each skill set involved. Where the same skill sets produce several deliverables, several deliverables can be represented by one participant. Choose your participants carefully so that every aspect of the project is covered without overcrowding.

Ensure you cover a wide range of levels from your organizational hierarchy. Senior managers often have a broad, interconnected, and long-range view that allows them to better identify strategic level risks. Conversely, junior participants (e.g., subject matter experts) often have a narrow, detailed view that allows them to better identify specific risks. Both views will be required to provide a comprehensive identification of your project's risks exposure.

There is no magic number of participants, but I find workshop effectiveness starts to decline as the number of workshop participants breaches fifteen participants.

This is in part because there is an inverse relationship between the number of people in a workshop and the willingness of people to speak out. In other words, people's inclination to contribute goes down as the number of participants in the room goes up.

People typically become less inclined to speak in front of a large group. And, they often incorrectly assume that in a large group someone else would have already covered any idea they might have thought of.

It's up to you, but I recommend keeping your number of workshops attendees below fifteen if possible, and definitely fewer than twenty.

4.2.1.2 Invitations

The second step is to send your invitations. Send your workshop invitations early to give yourself the best chance of getting the right people in the room. Even if you don't have all the required information, issue a save-the-date invitation so participants have a placeholder in their diaries. In the invitation, let the participants know that you'll be following up with an agenda and meeting material closer to the time of the workshop to ensure the material is current.

Scheduling the session well in advance allows the invitees time to juggle their schedules and make room for the workshop. But, no matter how far in advance you send the invitation, it is rare to find a time that suits everyone.

In the event of a conflict, it is usually easier to ask the more junior participants if they could reschedule conflicting appointments. If that doesn't solve the conflicts, senior managers often have personal assistants who can be invaluable in assisting with schedule conflicts. If you are still having difficulty, you can always ask participants to delegate their attendance to a subordinate or colleague who can cover the relevant subject area.

Ensure that you book enough time for the workshop to cover all the topics. If you book too little time, it can be difficult to reconvene a large group in a reasonable time. And, if rebooking results in a substantial time lapse before you can come together again, the loss of momentum can make it very difficult to pick up where you left off. If you have to book a follow up meeting, you will find that you will lose some momentum, even if you reconvene as early as the next day.

If there is a month's delay before you reconvene, it is nearly impossible to pick up where you left off. Take the precaution of booking too much time rather than too little time. People rarely begrudge being released early from a workshop. Conversely, running late in a workshop is almost universally reviled.

People can be productive in a workshop for only a couple of hours at a time, and not more than four to six hours per day. Keeping people in a workshop for any longer is a waste of their time and effort and is likely to reduce the quality of the outcome.

If the list of topics you need to cover is long, consider scheduling the workshop over a couple of days. Look for a distinct separation of subject matter to divide between the two days to minimize the loss of momentum from holding separate sessions.

Don't plan to work for more than two hours without a break. If you are planning to run a long workshop, it's a good idea to arrange refreshments for breaks and lunch. Plan to keep your participants rested and refreshed to maximize their contribution. Include the timeline for each of the segments in your invitation so that people know what to expect.

4.2.1.3 Agenda

The third step is to create an agenda. A good agenda states the objective, the activities, the timeline and the topics to be covered. It also provides context for the workshop.

Devise the workshop structure that you want to use and outline it in sufficient detail when you create the agenda. Workshop participants are much easier to manage when they've been informed of expectations beforehand.

Conversely, participants who have not been informed of what is expected can quickly derail your workshop. You will find it much easier to facilitate the workshop if you have a thoroughly thought-out and well communicated agenda structure to lean on.

The size of your project, the time available, and the perceived importance of your project will influence how you decide to structure your workshop. Depending on how you choose to structure the workshop, the agenda process steps may include

1. Risk identification – brainstorming to identify risks
2. Risk refinement – the refining of risk descriptions to an actionable level
3. Qualitative risk analysis – analyze and rank the risks according to
    - The probability of the risk event occurring
    - The impact to the project if the risk occurs
4. Plan risk response – comparing risk analysis results with risk criteria to determine risk response plans.[21]

Inform your participants of the workshop structure before the workshop to enable them to make a more valuable contribution to the risk workshop. If you don't have all the information when you send out the "save the date" invitation, you can always follow it up with an information-laden update later.

Some people split risk workshops into subgroups to divide up the work. They split into groups to speed up the process to ensure the workshop completes on time. I advise against splitting into groups. In my view, this approach dilutes the synergistic effectiveness of having many varied minds consider each risk.

[21] **Note**: one risk response might serve more than one risk, so the number of risk response plans may be less than the number of risks.

I have excluded quantitative risk analysis from the risk identification workshop. Quantitative analysis provides an objective analysis on risks and on the overall impact on the project. As this is an objective activity, there is little benefit in doing it in a workshop.

Often, risks that your participants propose while addressing one aspect of the project spark thoughts in team members about unrelated areas of the project. I try to maximize opportunities for these synergistic *idea triggers* by keeping the participants all in one group.

There are typically four distinct segments to a risk workshop (identify, refine, analyze, and plan response). As will be expanded on later in this chapter, each segment requires participants to be in a different mindset. It is beneficial to maintain the separation between segments rather than run each risk through all four segments, one risk at a time.

In my experience, it is more effective to run all the risks systematically through one segment at a time before moving onto the next segment.

This approach entails identifying all the risks at the outset. Once you have finished identifying the risks, then you would refine all the risks. Once you have refined all the risks, then you would qualitatively analyze all the risks. After they have all been qualitatively analyzed, then you would plan risk responses for all the risks.

This sequential approach to your risk workshop will help your participants to maintain the most appropriate mindset for each of the process steps.

4.2.1.4 The High-Level Workshop Approach

As humans, we are naturally inclined towards linear thinking. Start at the beginning, work through the middle, and then finish up at the end. This approach has its place. But, given the difficulties in cognitive shifting caused by mindset bias, your risk workshop is not one of those places.

In my experience, it is more effective to work through each segment in its entirety before moving on to the next. Spend time at the beginning identifying as many risks as possible. After you've finished identifying risks, then move on to refining all the risks. After all the risks have been refined, then analyze and rank all the risks and, finally, plan risk responses for all the risks. The high-level risk workshop process is depicted in Diagram 11, below.

Diagram 11 - Workshop Approach Process

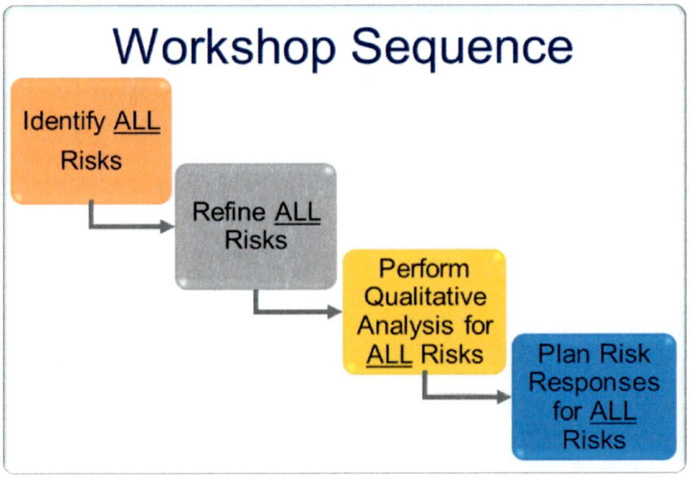

1. Identify risks – this is the actual discovery, initial description and recording of a risk

2. Refine risks – collate risks that overlap, amalgamate risks that could be addressed through a common risk response plan and dissect risks that can be logically broken down into separate risks with distinct risk response plans
3. Qualitatively analyze risks – assess the impact and probability and apply a risk score
4. Plan risk response – create a plan of an action to address every risk

Once a risk has been identified, participants may be tempted to progress to refine, analyze, and then plan a risk response for that individual risk. Although there is a logical progression in this approach, I would steer your participants away from it. The stages of the process require participants to be in different mindsets and moving from mindset to mindset incurs a switch cost.

Risk identification requires participants' minds to be thinking broadly. Participants should maintain an expansive and open focus during risk identification. This type of blue-sky thinking differs from the type of thinking required for the rest of the risk management process.

Refining and analyzing risks each require a more concentrated focus and analytic mindset. In my experience, most people can move relatively easily from the broad to the focused mindset. But they have trouble moving back to the broad mindset if they are asked to resume brainstorming.

Risk response plans requires a hybrid mindset. Blue-sky thinking is required to come up with innovative options for risk response plans, and analytic thinking is required to analyze and assess risk response plans to determine the most appropriate plan for each risk.

If you identify, refine, analyze, and then create a risk response plan for each risk as it comes up, your participants will have to change mindsets three times for each risk. This constant change of mindset is going to incur a switch cost that is almost certain to degrade the quality of your workshop outcomes.

Risk identification requires a blue-sky mindset in which the mind is unfettered and unrestrained. Do not impede the quality of your workshop output by forcing your participants to change their mindsets too frequently.

Of course, you don't want to constrain conversation or lose any additional risk information that might be provided during risk identified. If your participants move beyond identification, I recommend you record the information and then gently guide the group back to identifying all the possible risks at the outset while people are still in the broad blue-sky mindset.

During the risk identification phase, you should plan to encourage your participants to throw as many ideas as possible into the ring before assessing those risks. When it comes to risk identification, quantity trumps quality. It is better to collect as many ideas as possible, no matter how unrefined they are. You can refine the risks and improve quality later, after all the new suggestions have been exhausted.

Ensure that you communicate your approach to your participants when you send them the invite.

4.2.1.5 Pre-reading Material

The fourth step in the preparation process is providing the pre-reading material.

Ideally, you will send your participants the preparation material early enough so that they get a chance to read it, but late enough so that it's not out-of-date by the time you conduct your workshop.

Participants should get the material early enough so that they have an opportunity to allow the information to percolate through their thinking. But, it has to be sent close enough to the workshop date, so they don't base their preparation on information that is no longer relevant.

A fundamental element of an effective workshop is participant preparation. Participants are more likely to provide effective input when they are well informed by the time of the workshop. The facilitator should send the meeting purpose, agenda, and relevant documents in advance.

Sending this information in advance enables invitees to prepare ahead of time, allowing them to hit the ground running on the day of the workshop.

Another benefit of providing information early is that you'll give those who prefer to analyze ideas thoroughly, before they speak about them, time to marshal their thoughts beforehand. They will be much more likely to contribute to the discussion when they've had time to think through the information prior to the workshop.

Prior to your pre-reading, your invitees will have varying levels of information about your project. You need to provide a broad background on the project so that the stakeholders who are less familiar with your project can be effective in your risk workshop. Your pre-reading should level the playing field and provide a more common level of project knowledge.

Accordingly, you need to send them the appropriate content. The key word here is *appropriate*. They need an understanding of the project goals, scope, and challenges, but you don't want to overload them with information.

Providing participants with too much reading material will backfire. They will either not read it, or they will resent having to wade through the excess.

I recall one risk workshop where the PM sent a ninety-page business case and an eighty-page slide pack to read in preparation. Not only was it time-consuming, but it also felt like the PM couldn't be bothered filtering the appropriate information. It was as if she had tossed a manual on my desk and said, "The information you need is in there somewhere. Go find it!"

If you want your participants to read the information you send them, you need to ensure the information is targeted and not excessive. You need to send your invitees curated information. The likelihood of a participant reading the pre-workshop information and the amount of the pre-reading information you send has an inverse relationship.

In other words, the more reading material you send, the less likely participants are to read it. The less reading material you send, the more likely participants are to read it. The concept is depicted in Diagram 12, below.

Diagram 12 - Amount of Reading Material You Send vs. the Likelihood of Your Stakeholders Reading It

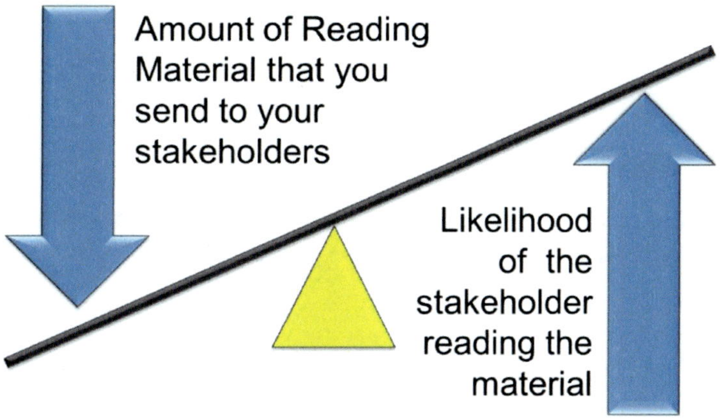

At a minimum, the information you provide before the workshop should include

- A description of the major deliverables of the project
- Project goals
- Project objectives
- Project milestones
- Project dependencies
- Corporate calendar
- Project scope
- Project assumptions
- Project constraints
- Procurement strategy
- Conceptual design

Also consider sending excerpts of the following documents

- Cost estimates
- Duration estimates
- Issue log
- Lessons learned register from similar projects
- Requirements documentation
- Resource requirements
- Stakeholder register
- Agreements

- Procurement documentation
- Any relevant academic studies
- Benchmarking reports
- Industry studies of similar projects

4.2.1.6 Critical Path Method

An analysis of the critical path focusses attention on the tasks that are most susceptible to schedule slippage, as they don't have any float. It focusses attention on the tasks that are most likely to be associated with schedule risk.

> *"A project without a critical path is like a ship without a rudder."*
>
> D. Meyer, American Lawyer

According to Wikipedia: "In project management, a **critical path** is the sequence of project network activities which add up to the longest overall duration, regardless if that longest duration has float or not. This determines the shortest time possible to complete the project."

A **near-critical path** is the sequence of project network activities that *would become* the next critical path if there were a "slippage" in the current critical path that is beyond the available float.

If you have mapped your critical path for your project, it can be a useful tool to use in your workshop. Obviously, if you're too early in your project to have developed one or are using a methodology that doesn't typically use a critical path (I'm looking at you Agile!), then you would skip this tool.

The critical path is rarely worked out manually anymore, outside of the classroom. Project management software does it for us. If you're using project software such as MS Project or Primavera P6, you can use the software to view the critical path.

Both of these software programs also allow you to apply a filter that allows you to set the parameters to display near-critical path activities as well. A near-critical path can be important, because it shows the tolerances for critical path risks before near-critical path risks become relevant.

A high-level project schedule showing the critical path (and near-critical paths) and showing high-level activities can be a useful tool for your workshop. A schedule with the critical path highlights the activities that would immediately affect the project schedule if they were impacted by a risk. And, a schedule showing the near-critical path highlights the tasks that "would soon affect the project schedule" if a critical path activity slipped past its float.

If you have a defined project schedule with the critical path and near-critical path, it is useful to send it as part of the pre-reading.

4.2.1.7 Objectives, goals, and benefits

Objectives, goals, and benefits define the direction of the project. They are the attributes that will be used to determine your project's success. Anything that may interfere with achieving them will be a risk. If you have developed one, a benefits dependency network diagram (BDN) can be a great visualization tool to help communicate how the different aspects of the project interact in achieving project benefits.

If you have defined objectives, goals, and benefits (or a BDN), it is useful to send them as part of the pre-reading.

4.2.1.8 Project milestones, a project schedule, and a corporate schedule

Project milestones, a project schedule, and a corporate calendar will provide time references that you can use to prompt the identification of time-dependent risks. An example of a time dependent risk would be a risk that there will be reduced availability of finance stakeholders at the end of the financial year.

If you have defined project milestones, a project schedule, and a corporate calendar, it is useful to send them as part of the pre-reading.

4.2.1.9 Deliverables and scope

Deliverables and scope define the parameters in which your project operates. You can use them in the workshop as a prompt to probe areas that have not been explored. Deliverables and scope can also be helpful to bring people back to topic if the conversation heads off on a tangent.

Scope defines the sandbox you are playing in. Everything inside the sandbox represents all that is in scope for your project. Everything outside the sandbox represents all that is out of scope for your project. You can use scope to bring people back into the sandbox if they start playing outside of it, and you can use it to ensure you explore every corner of the sandbox.

If you have defined deliverables and scope statements, it is useful to send them as part of the pre-reading. It is equally helpful to share any scope statements that define what is out of scope for your project.

4.2.1.10 Dependencies, assumptions, and constraints

Dependencies, assumptions, and constraints can also stimulate your participants to identify risks caused by external factors. Providing this information to your participants before the workshop will allow them to consider the related areas of uncertainty.

When you conduct the workshop, consider each of the **dependencies** one by one. This will give the group an opportunity to examine the factors that the project is reliant on. Exploring these factors may unearth associated risks.

Examining the **constraints** will help identify any risks associated with the restrictions on the project. Similarly, testing the **assumptions** will identify any risks relating to the suppositions that have been applied to the project.

Dependencies, assumptions, and constraints often trip up PMs, because they're not treated with the same rigor as risks.

In most organizations, there is a set process for managing risks. Risks are identified, the probability of the risk occurring is estimated, and the impact the risk would have on the project is appraised. A risk score is calculated, a risk response plan is agreed, and the risk is regularly reported on and reviewed.

Unfortunately, organizations rarely have such mature processes for managing dependencies, assumptions, and constraints. And, even if they are reported on, they rarely attract the same level of attention as risks. They are rarely quantified and, even if they are reported, they rarely get more than a cursory glance by management.

A few years ago, I worked with a project manager who had drawn up a list of critical assumptions. When I suggested that she reword them and categorize them as risks as well, she dismissed the idea saying that they were already covered as assumptions. Six months later, she found out that three of those major assumptions were erroneous. Those three assumptions were going to cause the project budget to double and delay delivery by eight months.

Even though she had reported her assumptions, management hadn't taken any notice of them. Management was focusing on the risks. They were shocked when they heard the bad news. They lost faith in the PM's ability to deliver the project. She was unfairly relieved of her project and was asked to leave the organization.

The PM had correctly identified the assumptions. She had correctly reported them as assumptions, but she had not highlighted to management the risk that lay within those assumptions.

Management didn't understand the probability, nor the impact of the risks embedded in those assumptions. They didn't understand what would happen to the project if those assumptions were proven wrong. Management overlooked the potential impact to the project, and they were caught unawares when the budget was blown, and the schedule unraveled.

It is imperative that you plan to review anything that has been identified as a dependency, assumption, or constraint. There is always risk lurking within dependencies, assumptions, and constraints. It is most often just a case of rewording them to tease out the risk that lies within.

Provide your invitees with dependency, assumption, and constraint information before the workshop.

4.2.1.11 Presentation

A presentation slide pack gives participants something to focus on throughout the workshop. Presentation packs are especially useful for introducing the workshop. I like to use MS PowerPoint presentations, as they allow you to easily incorporate reference materials, such as diagrams, schedules, and lists, as you speak on the topics.

If presenting on-screen is not practical, you can provide a printed copy of a presentation. But, providing a printed copy is not ideal. The disadvantage of providing a printed copy is that you lose some control. Participants will be tempted to flip ahead in the slide pack, reading ahead instead of paying attention to what you are saying.

Whatever method you decide to use, it helps to follow the chronological sequence of events your workshop will take and to tie the presentation to the key points you are trying to make.

Using pictures, diagrams, or images that reinforce the points you are speaking about can be highly effective. Using visuals instead of text prevents people from simply reading the words on a screen instead of listening to what you're saying.

Presentations are most effectively used in workshops to show visuals that reinforce your key points, and presentations allow you to provide complex information that is too complicated or interconnected to convey with speech alone.

Take the example of trying to talk through a schedule. Speaking about the tasks' start dates, end dates, dependencies, and resource utilizations would be almost impossible to do with words alone. It is doubtful your participants would be able to follow your explanation if you relied on spoken words alone.

A visual, such as a Gantt chart, depicts complex schedule information with reassuring economy and simplicity, and they can be incorporated into your presentation. A Gantt chart is a good example of a visualization that can provide a helpful reference while you talk your audience through the activities on the schedule.

I don't recommend sharing your presentation before the workshop, but it's something that you'll want to develop and refine before the workshop.

4.2.1.12 Workshop Material

There are some materials that you can prepare in advance to ease your facilitation of your workshop.

It is a good idea to equip your participants with sticky notes (aka Post-it notes) to capture risk descriptions (see Section 4.5.1 for how sticky notes are used). Provide each participant with a pad of sticky notes and a pen for writing. Even though most will bring their pens, you want to make sure your workshop is uninterrupted, so it is wise to bring enough spares.

In my experience, it is too difficult to both facilitate the workshop and record all the notes. Rather than doing both poorly, I arrange for someone to act as a scribe and type notes straight into the software I'm using (ergo, you would need both a scribe and a laptop). Otherwise, the tried and tested method of writing notes out longhand and transcribing later works too (ergo, you will need a scribe with a pen and a notebook).

As mentioned earlier, I prefer to use a live slideshow presentation for the introduction as it keeps people on track. A live slideshow also prevents the loss of attention that occurs when your participants start flipping pages instead of paying attention to the slide you are speaking about. Clearly, a live presentation requires a projector with a screen or a large screen TV.

Ideally, the room should also have a whiteboard. A whiteboard can be useful for posting risk categories. A whiteboard can also serve as a "parking lot" (a place to record items that are out of scope). An easel with a poster-sized paper pad can be used for the parking lot if no whiteboard is available.

In preparation, I also find it helpful to prepare oversized printouts of the risk categories in huge fonts to act as category headings. 42 example categories are listed in Appendix A, or you could pick up our complete set of *42 printable category headings* at www.ictrisk.com.

Prior to the brainstorming, stick your category headings on the whiteboard or a wall of the room so they can be used as categories for any sticky notes tacked below them.

When the time comes, participants can put up their sticky notes under the relevant category headings. Multiple whiteboards can be helpful for this task, as can a smooth wall that sticky notes will stick to.

It can be beneficial to gather additional material that can provoke thought about risk identification. As well as the tools outlined in this book, you might include the risk register from a previous project, an industry checklist, or a list of risk events taken from a previous project's "lessons learned." These can stimulate thought and discussion in the participants. Consider gathering this material well in advance.

4.2.1.13 Facilities

Take into consideration the size and layout of the room in which you will host your workshop. Participants should be comfortable, which means offering adequate seating, access to a writing area, and enough room to spread out.

I once participated in a workshop that had so many people crammed into a small room that half the participants were forced to stand throughout. It was unpleasant, and I am sure I wasn't the only one who couldn't wait to get out. You don't want people holding back risks because they're uncomfortable and just want to get out of the room as soon as possible!

Although I have successfully used video conferencing to conduct risk workshops over multiple locations, it's not easy. Successful video conferencing depends on the facilities you have available, and using VC adds complexity and stress. It can be difficult to position a whiteboard so that all participants can see and having people in different locations impedes the flow of ideas in a workshop.

It is more effective to have all your participants in the same location if possible. Use video conferencing only as a last resort.

4.3 Running the Workshop

Now that you have planned how you're going to run the workshop and have fully prepared for it; it is time to execute. Facilitating a risk workshop follows a series of steps. These steps are

1. Initiate workshop
2. Identify risks
3. Refine risks
4. Perform qualitative analysis on the risks
5. Develop risk response plans

4.3.1 Initiate the Workshop

Once your attendees have settled into the room, it is time to initiate the workshop.

Instead of diving straight into risk identification, spend a bit of time preparing your participants for what is to come. Remind them of why they're there, what is expected of them, and how the workshop is going to be run. If you're going to get the most complete set of risks possible, you need to set the context for the project and create a common understanding of risks.

The steps I recommend for initiating a risk workshop are

1. Set the context of the project
2. Set the context of the workshop
3. Introduce workshop methodology
4. Introduce the workshop roles and expectations
5. Describe risk management methodology
6. Introduce the agenda
7. Define terms and introduce risk levels

You will have a much easier time facilitating the workshop if your participants have a common understanding of what the workshop objective is, how the workshop will be run, and what is expected of them.

4.3.2 Set the Context of the Project

After you have welcomed your participants and gone through the introductions, it is best practice to set some context for the project. You have already provided much of this in the pre-reading, but the project may have moved on in the interim and some of your participants may not have read the pre-reading.

In addition to reiterating the information provided in the pre-reading material, provide a quick overview of the project. In your context-setting, you should include the following:

- An overview of the project
- Project goals
- Project objectives
- Project deliverables
- Project milestones
- Project assumptions
- Project dependencies
- Project constraints
- High-level project schedule

If you have any scoping artefacts that could help draw the project scope boundaries, it would be useful to share them at this point. Even if it is just a simple table showing what is *in scope* and what is *out of scope*, it would be useful to share.

Sharing these artefacts with your participants will ensure your participants' thought trains stop at all *in scope* stations. And, sharing these artefacts will derail any thought trains heading to a destination *out of scope*.

I like to keep any scoping diagrams on a wall so that I can refer to them while facilitating the workshop. When a discussion goes out of scope, you can simply reference the diagram and easily redirect the conversation back to risks that are in scope.

4.3.3 Set the Context of the Workshop

Once you've set the context for the *project*, the next logical step is to set the context for the *workshop*. The participants now have the details of what the project aims to achieve, how you intend to do it, and where you are in the process of delivering the project. Now is the time to set the context of the workshop. Ensure that everyone is aware of

- The purpose of the workshop
- The objective of the workshop
- The outcome of the workshop
- The outputs of the workshop

4.3.4 Introduce Workshop Methodology

As with any meeting, it is good practice to set expectations at the outset. Introducing the workshop methodology will ensure that your participants know what is coming and will reduce any resistance as you run through the process.

4.3.5 Workshop Rules

Any time that you bring together a large group of people for a workshop, there is a danger that the workshop will slip out of your control. Therefore, it is important to establish rules at the outset. Setting the rules at the beginning will give you the best chance to keep the workshop on track.

You or your organization may have a standard set of workshop rules. It is good to use these, but for risk identification, it is imperative to also add a rule. Add a rule against criticism.

Criticism is poison for blue-sky thinking. And blue-sky thinking is essential for risk identification. Risk workshops are most effective when they create a safe environment that encourages collective brainstorming for risk identification.

Collective brainstorming is a group effort that aims to generate output that is greater than what could have been achieved by the sum of the individual efforts. Criticism must be banned to get the most out of collective brainstorming.

Collective brainstorming allows the ideas put forth by one participant to stimulate thought processes in other participants. Those other participants can in turn contribute to the idea and can trigger related or unrelated ideas. Collective brainstorming is about collecting as many ideas as possible and developing those ideas collaboratively.

Criticism should be banned, because the risk identification stage should focus on quantity. Quality should be ignored during risk identification. Your objective at this stage is to get as many ideas on the table as possible. You will get the chance to screen for quality later.

By deferring criticism and by encouraging out-of-the-box thinking, you can create the type of safe environment in which brainstorming can really flourish. By deferring criticism and by encouraging out-of-the-box thinking, you reduce your participants' inhibitions, stimulate idea generation, and increase the aggregate creativity of the group.

You will likely have participants who are inclined to refine, criticize, or strike down ideas as soon as they are suggested. If you let them follow their inclinations, they will greatly reduce the value you get from brainstorming. If you let them, they will choke off the number of ideas that your participants suggest.

Make sure you clearly state the reason for the brainstorming rules and assure your participants that you are taking an iterative approach. In other words, make it clear that they will have an opportunity to refine the ideas later in the session.

It can be difficult to enforce rules, especially if there are senior managers or executives as participants. To maximize compliance to rules, explain the reasons for the rules and ask the participants to agree to them at the outset of the workshop.

This tactic can be especially effective if you single out the most senior people in the room and ask them to agree to the rules. If the most senior people agree to your workshop rules, the rest are much more likely to fall into line.

It is much easier to maintain compliance to rules during the workshop if the group has agreed to abide by them at the outset of the workshop.

Helpful rules include

- No interruptions – consider an off-site location to minimize the chance of interruptions by other staff
- No judgments or criticisms – during risk identification, everything gets recorded; refinement comes later
- All to participate
- No mobile/cell phones

4.3.6 Introduce Meeting Roles and Expectations

Describe the workshop roles so that all participants know what to expect. Typically, there are three roles in a risk identification workshop, although this is not written in stone. These roles and their responsibilities are normally

Facilitator

- Prepare carefully to anticipate the contents, the politics, and the likely breakdowns of risk

- Provide the introduction, set the context, present the agenda, and propose the workshop rules
- Provide common definitions of terms and approach
- Provide explanations and examples of risk syntax (risk meta-language)
- Inform participants of the target "level" of risks being sought and provide examples
- Use tools and techniques to help people probe for risks
- Manage tangents and rein-in discussions that go out of scope
- Probe areas of scope that may flush out more risks
- Manage and steer discussions – keep the level and direction of the workshop generally on course without constraining the flow of ideas or choking off beneficial discourse
- Prod and stimulate thought in diverse areas of uncertainty
- Ensure all items on the agenda are covered

Note Taker

- Record risk descriptions clearly and precisely and aggregate/group risks to the most appropriate level
- Concentrate on getting all the points down with the intention to write the notes up and organize the structure later

Participant

- Concentrate on uncertainty – think through all aspects of the project and then explore all other conceivable "what ifs"
- Explore both threats AND opportunities

4.3.7 Describe the Risk Management Methodology

The PMBOK 6th edition lists seven risk management processes. The seven processes are

1. Plan Risk Management
2. Identify Risks
3. Perform Qualitative Risk Analysis
4. Perform Quantitative Risk Analysis
5. Plan Risk Responses
6. Implement Risk Responses
7. Monitor Risks

This book covers all seven processes but emphasizes four of those processes.

The primary focus of this book is on the identify risks, the perform qualitative analysis process, the perform quantitative analysis process, and the plan risk responses process.

In this book, I have followed the processes outlined in the PMBOK, but I have decomposed parts of the process, where I thought the reader would benefit from a lower level of detail. This decomposition is amplified in the focus areas.

Providing your participants with an outline of the risk management process will forestall questions as it will include a description of what will be addressed in later stages of the process.

Many of your participants may already be aware of the risk management process, but it is good practice to do a high-level run through of the risk management methodology. Doing so further sets the context for participants and ensures a common understanding among the group.

Risk management processes include similar steps to the list below. It is best to run through the methodology that your organization uses. The list below is an example only.

1. Identify risk
2. Refine risk
3. Analyze risk
4. Determine a risk response plan

These steps have been depicted in Diagram 13, below.

Diagram 13 - Workshop Process Diagram

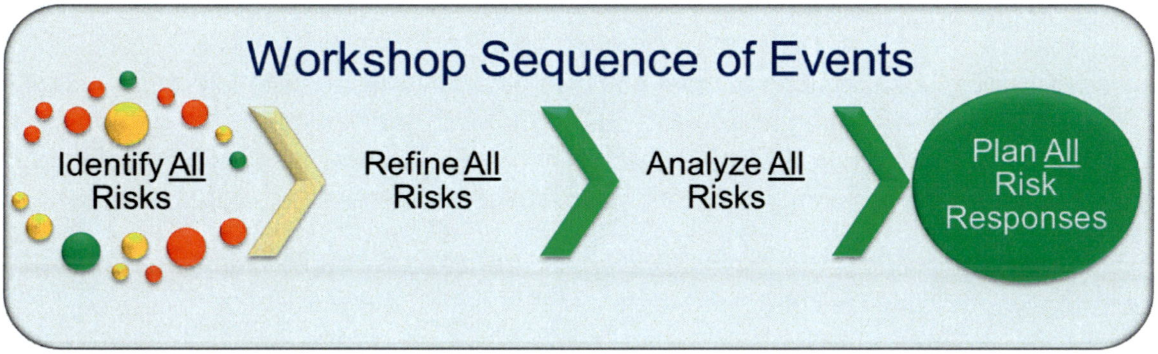

As risk identification starts near the beginning of the risk management process and provides the source material for the rest of the process, I cannot overstate the importance of getting as much down as you can from the outset. Risk identification is the heart that pumps blood through the entire risk management process.

The whole project can be impacted if you miss risks, so ensure you highlight the importance of being thorough to your participants.

4.3.8 Introduce the Agenda

People generally don't like surprises. Provide them with as much Information as is practical during the introduction of the workshop agenda. This will give them total transparency and will prevent surprise during the workshop.

As with the setting of workshop rules, you will have an easier time gaining attendee "buy-in" during the workshop if you set expectations at the outset. If you ask participants to agree to adhere to your workshop agenda, you will have an easier time achieving your workshop objectives.

During the workshop, you'll find people are much more inclined to follow an agenda that they have agreed to. People have a natural tendency to want to remain consistent with what they have said. If they said that they will adhere to your methodology at the outset, they are much more likely to do so throughout the workshop. You'll find this invaluable when "pulling in" wayward conversation and when managing your agenda.

This concept of asking for an agreement to your workshop methodology ties in with the concept of workshop rules. But, it also gives you a chance to show what tools and techniques you will use in the workshop. And, it gives you a chance to communicate the actions you will take if participants deviate from the methodology.

Separating the request to agree to the workshop agenda from the request to adhere to the workshop rules reduces participants' resistance.

It is easier for people to accept being asked to do "X" and then later being asked to do "Y" than to be asked to do "X + Y" at the outset. People are much more likely to agree to a second request if it aligns to an earlier request that they have already agreed to.[22] We all have a natural urge to maintain consistency in our behavior.

As discussed earlier, brainstorming is most effective when criticism is withheld and the contribution of any and all ideas is encouraged. When you inform your participants at the beginning that you welcome *any* ideas, you encourage participation. When you tell them that you will not tolerate criticism during the brainstorming session, you further break down the barriers to engagement.

And, when you state at the beginning that you'll be calling people out if they criticize while brainstorming, you will find less of a need to do just that during the workshop.

It's one thing to ask that everyone take part, but actually convincing all your participants to contribute is quite another thing. For the various reasons described in section 4.1, people often resist contributing or will only contribute the risks that they feel secure in proposing.

Highlight to your participants that all their ideas will get airtime and that there is no harm in voicing ideas. Declare the workshop to be democratic and explain the reasoning behind it. This declaration will increase your workshop participation rate.

[22] Robert B. Cialdini – Influence: *The Psychology of Persuasion* – Chapter 3

Review the agenda with your participants to provide an opportunity to highlight the sequence that the risk identification workshop will follow.

4.3.9 Define Terms

Defining terms at the start of the workshop will provide the participants with a common understanding. A common understanding will save time throughout the workshop.

It's good practice to keep the definitions of terms visible so that you and the participants can refer to them as needed. I like to print the definitions out beforehand and post them on the wall before the workshop starts. That way they can be easily referred to during the workshop.

There are a number of aspects that are encompassed in "defining terms" including

1. Defining the meaning of words and phrases
2. Defining the syntax of describing risks
3. Defining the most appropriate level of describing risks
4. Defining the types of risk that are worth capturing

4.3.9.1 The Meaning of the Words and Phrases

It's easy to assume that terms such as *risk*, *issue*, *cause*, *impact*, *probability,* and *risk response plan* are common enough that people would all understand their meaning.

Time and time again, I have seen that assumption proven wrong. Even when everyone thinks they understand the definitions of the terms, often those understandings differ from one individual to the next.

Define the terms of risk lexicon at the outset of your workshop and provide examples for each. This will save time and increase the chances of reaching your workshop objectives.

Hopefully, your organization has a set of defined terms that you can share with the group to ensure a common understanding. If not, you will have to provide definitions of terms.

At a minimum, define terms such as *risk*, *cause*, *impact*, *probability*, *risk response plan,* and *risk threshold*. If your organization doesn't have standard definitions, feel free to base your definitions on those in the glossary of this book.

It's also worth noting that using the word *risk* tends to narrow thinking and skews people's thinking towards the negative. It is becoming more and more common in the risk management industry to use terms such as *areas of uncertainty* in risk workshops. This is especially true if your aim is to encourage broader thinking and to incorporate the consideration of *positive risks* (i.e., opportunities).

When facilitating the workshop, word choice can have a significant effect on the workshop output. A facilitator who talks about *identifying risks* can give the impression that there are tangible things out there that simply need to be discovered. This can lead people to discount the unknown unknowns.

Terms like *areas of uncertainty* do not have the same connotations, and they have been found to encourage people to become more inclined to explore the unknown unknowns.

As a facilitator, you don't want to reinforce the illusion that you already know everything that could possibly impact your project in the future!

4.3.9.2 The Syntax of Risks (Risk Metalanguage)

The *syntax* of a risk simply refers to how you order the words describing risk. If your organization has a standard for risk syntax then, naturally, use that. If not, both the *Project Management Institute* and *Axelos Management of Risk* use a common syntax for describing risks.

Their risk syntax follows the basic formula: Cause(s) + Risk Event + Effect(s) on Project Objective(s). There are four examples below.

> "A member of the development team suffers an extended illness (cause), resulting in less code written per day (risk event), which leads to the schedule slipping (effect on *schedule*).
>
> "As a result of using new hardware (cause), unexpected integration errors may occur (risk event), which would lead to budget overruns on the project (effect on *costs*)."
>
> "Because our organization has never done a project like this before (cause), we could misunderstand the customer's requirement (risk event) and build a solution that does not meet the customer's performance criteria (effect on *quality*)."
>
> "We have decided to outsource production (cause); we may be able to learn new practices from our third-party partner (risk event), resulting in increased productivity and profitability (effect on *benefits*)."

Using a standardized risk syntax ensures that risk identification recognizes the actual risks. Without the use of standardized risk syntax, the risk descriptions can produce confusion and distraction later in the risk management process.

The cause-event-effect relationship is not necessarily a one-to-one-to-one relationship. An event can have multiple causes. And an event can have multiple effects.

The fact that risks can have multiple causes and/or multiple effects is an important point to raise with your participants when considering risk syntax. You will need to agree to the method you will use to record risks with multiple causes and/or effects.

When I record risks, I prefer to split up risks that have multiple causes and/or effects, as I find it easier to manage them later in the process.

For example, due to the resource-heavy nature of ICT projects, we often find risks whose effects influence both costs and schedules. Even though such risks might have the same cause and risk events, I prefer to record them as separate risks.

Let's consider our previous example of a team member falling ill. If we assume that we pay our developers sick leave, we can record it as two risks to reflect the different effects, as shown below.

> Schedule risk: A member of the development team suffers an extended illness (cause), resulting in less code written per day (event), which leads to the schedule slipping (effect)

Cost risk: A member of the development team suffers an extended illness (cause), resulting in less code written per day (event), which leads to increased costs (effect)

Recording these two risks separately allows you to categorize them and report on them more accurately.

It is up to you to decide how you record risks that have multiple causes and/or multiple effects. Whichever method you decide to use to record such risks, let your participants know at the outset how you intend to record and manage multiple causes and multiple effects.

4.3.9.3 Risk Levels

You need to define risks at the most appropriate level if you are to manage them effectively. You need to define risks at a level that allows you to apply meaningful risk response plans. And you need to define risks at the most appropriate level if you are to manage them effectively. You need to define risks at a level that prevents them from slipping under a threshold and simply being added to the "watch list."

If you define risks too high, you will not be able to apply meaningful risk response plans to the risks. For example, consider a risk defined as "Stakeholders are too occupied with other corporate activities to contribute to project tasks when required, causing the schedule to slip." You may find you have described the risk too high to apply a meaningful risk response plan.

Conversely, if you define a risk at a level that is too low, it may pass under a threshold and simply be added to the watch list. For example, consider if you have a risk defined as "The technical superintendent will be occupied with critical operational activities rendering him unavailable to review the solution design document causing the schedule to slip."

Such a risk would be too granular. The risk has been defined about a single role being available to review a single document. The risk score for such a granular risk would be too low to receive attention. A risk defined at this level would likely sit beside several other similar risks (i.e., for all the SMEs required to review the solution design document and other documents).

Individually, those risks might have low risk scores, but when you aggregate those risks together, the risk score rises, and you have a combined risk that will be assessed as important enough to warrant an active risk response plan.

There will be a *Goldilocks zone* for defining risk. This will be a level that is neither too high, nor too low. It will be a level that is "just right." A level where the risks are at a low enough level so that you can assign meaningful actions to them and at a level high enough that you will not have granular (low-level) risks slipping under a threshold and added to the "watch list."

As a general rule, if two or more risks have the same risk response plan, then you should consider rolling them up into one risk.

For example, if you have several risks related to subject matter experts (SMEs) being available to review project documents, you may want to combine the risks. They might combine into one risk, such as "If operational demands are greater than expected, SMEs will not be available to perform the required reviews of documents on the critical path, resulting in a delay to the project schedule."

The concept of defining risks at the most appropriate level has been depicted in Diagram 14, below.

Diagram 14 - Risk Level Diagram

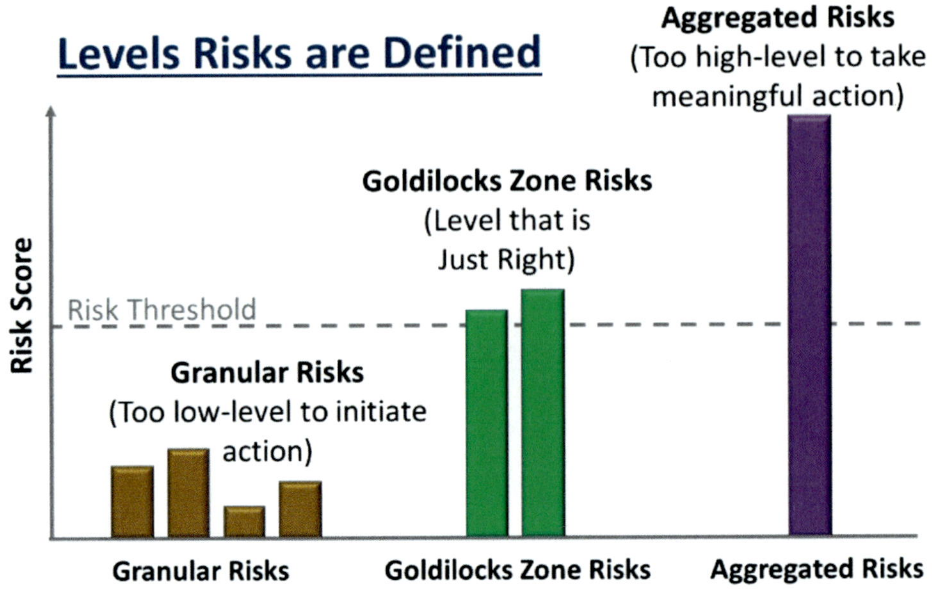

4.3.9.4 The Types of Risk Worth Capturing

Before you start the workshop, you should address the types of risk that should be in scope for your workshop. You need to capture the risks that are relevant and exclude anything that is irrelevant.

You should exclude risks that are so far-reaching that you couldn't conceivably have any other risk response plan than "accept."

Yes, a giant meteor might crash into the earth and disrupt your project. Yes, global nuclear war could break-out and impact your project. And yes, a new genetically designed superbug zombie pandemic could disrupt your project.

There is little point in recording such risks. These are extreme examples, but you will likely find some risks proposed that are similarly unhelpful. Outline at the outset of your workshop the types of risk that are worth capturing, and the types of risks that you can leave for the film industry.

4.3.9.5 Risks that have tasks already assigned to deal with them.

You should also exclude risks that already have tasks assigned to deal with them in the work breakdown structure (WBS). If the risk response plan for a risk is already a task on your project plan, then you want to consider carefully what risk actually remains given that a task has already been planned.

We should only identify and manage the risks that aren't covered by other project management processes. If risks are already covered off by other project management processes, such as a task on the schedule, then we would waste time in identifying and managing it as a risk.

There is a danger that everything can be described as a risk. We need to instruct our participants on the types of risk identification that is useful and the type of risk identification that is useless.

The aim of risk identification should be to define risks that you can sensibly apply a risk response plan to and to avoid recording risks whose risk response plan is already in the WBS. An example of an impractical risk is given below.

> **Project Objective:** Provide sales staff with access to sales information when unable to connect to the corporate network.
>
> **Risk**: Failure to provide sales staff with access to sales information when unable to connect to the corporate network.
>
> **Risk Response Plan**: Mitigate – implement tools to allow sales staff access to sales information when unable to connect to the corporate network. (Note that this is the stated objective of the project and has already been addressed in the WBS.)

Although it is easy to see that there is little benefit in listing risks that are already addressed by actions in the project plan, it is worth highlighting this for your participants at the start of the workshop to prevent that type of thinking.

4.3.9.6 External Broad-Reaching Risks

There will be some external risks that are so broad reaching that there will be no other risk response plan other than "accept." These risks should be recorded but there is no benefit in including such risks in your active risk management. They should be acknowledged, but there is no benefit in spending time and effort on them.

You may have some broad-reaching external risks that you can do little about. You might have a risk such as "the economy could head into a recession causing the product you are developing to lose appeal and the project to be cancelled."

Unless your circumstances enable you to hedge against the risk of a recession, you should acknowledge the risk and move on. If the risk is so broad that you can't conceivably do anything about it, then your efforts will be better spent elsewhere.

Any efforts beyond recording and reporting on broad-reaching external risks that have no possible risk response plan other than "accept" are a waste of time.

Sure, your project could be impacted by a range of external broad-reaching risks. But, as stated in section 4.3.9, *Types of Risks Worth Capturing,* if there's nothing you can do other than accept those risks, then don't waste time on them.

4.4 Risk Identification Methods

> *"It's easy to see, hard to foresee."*
>
> Benjamin Franklin, American polymath

Risk identification is both easy and hard. It's easy to see the obvious risks. Your participants will be quick to identify the risks that are common to most projects. Those are easy. But, your project will also have risks that are much harder to foresee. Those are the risks that you need to put effort into identifying. Those are the risks you need to approach from different angles, using distinct tools and techniques.

4.4.1 Interviews

Interviews do not provide the same rich context for risk identification that workshops do. As such, they cannot provide the chance for ideas to bounce up against each other and stimulate new lines of thought. But, they do still have a place in risk identification.

4.4.1.1 Senior Stakeholders

You will likely have at least one stakeholder who is too senior (read "busy") to be able to spare the time to attend a workshop. And you may have a senior stakeholder whose perspective is too broad to be of much benefit in a workshop. It is for those two reasons that interviews may be appropriate.

First, let's consider the busy stakeholders. Senior managers often have so many other demands on their time that they cannot dedicate hours for a risk workshop. The demands on their time are so extensive that they often divide their focus and multitask in any of the meetings that they do attend.

Even if they did attend your workshop, they would likely spend their time responding to emails or working on other tasks. They rarely have such a long period of time to dedicate to a workshop, and they are rarely inclined to dedicate the focus and attention that project risk identification requires.

Second, senior managers tend to have a more strategic viewpoint. A strategic viewpoint has value of its own, but it limits the benefit they can contribute to a workshop. It is impossible for senior managers to understand all the details of all the activities of the people who report to them.

Senior positions necessitate a higher-level perspective where they focus more on the strategic issues. Their focus will be at a different level from the rest of your participants. And that difference in focus limits the amount of synergy they could generate by bouncing ideas off others in a risk identification workshop.

For these reasons, it is best to schedule a short, focused interview with senior managers to ensure that you capture their concerns and uncertainties. Their strategic perspective will likely enable them to identify risks that may not occur to your other participants.

4.4.1.2 Specialist Subject Matter Experts

Another group of stakeholders who you may want to interview, instead of including in your workshop, are specialist subject matter experts. I have worked on projects where we engaged specialist subject matter experts to perform specific isolated tasks. These specialist subject matter experts are also candidates for interviews.

These specialist subject matter experts were expensive external resources who had little knowledge or interest in the project's overall objectives. They had little interaction with the project outside of their specific tasks. Including them in a workshop would yield little benefit and would be a poor use of their expertise.

Specialist subject matter experts may be able to identify risks specific to their area of expertise that no one else would be aware of. It is best to schedule an interview with these experts to ensure you capture their specific concerns in your risk identification. Due to their specialist and unique knowledge, you should try to solicit their estimate of the probability, impact, and suggested risk response plans for their specific risks all in one sitting.

4.4.2 Questionnaires

Depending on the nature of your project, you may have stakeholders who are located remotely. While you may be able to engage remote stakeholders in a workshop through video conferencing, their interaction is generally less than if they were physically present. You may want to supplement remote stakeholders' input with questionnaires before conducting the workshop.

If you have stakeholders spanning the globe, you may also have conflicting time zones that make it impossible to conduct a VC workshop that can include everyone. If time zones are a constraint for your project, you can use questionnaires to ensure that you get remote stakeholder's input into risk identification.

With both interviews and questionnaires, ensure that they are done before the workshop so that those stakeholders' contribution can feed into your risk workshop. You never know what new ideas their contributions may spark with your workshop participants.

4.4.3 Workshop

Axelos M_o_R introduces the clever idea of identifying risks through a two-step process. These steps are (1) using tools and techniques to set the context for identification and (2) using tools and techniques to identify risks.

I have expanded on these steps into their composite tools and techniques as below.

1. Set the context for identification via the following tools and techniques
   - Stakeholder analysis
   - PESTLE analysis
   - SWOT analysis
   - Horizon scanning
   - Lessons learned and previous risk registers
   - Pre-mortem
2. Identify risks via the following tools and techniques
   - Brainstorming (including Delphi technique and Nominal Group technique)
   - Root cause analysis

- Cause and effect analysis (Ishikawa diagrams)
- Project constraints
- Assumption analysis
- Constraints analysis
- External risks analysis
- Mind mapping and affinity diagramming
- Checklist analysis
- Business process analysis

As depicted below in Diagram 15, your attendees will have three levels of risk awareness. These are "obvious risks," "reserved risks," and "suppressed risks." The fourth level is for risks that they have no awareness of, "unimagined risks."

Diagram 15 - Levels of Risk Awareness Diagram

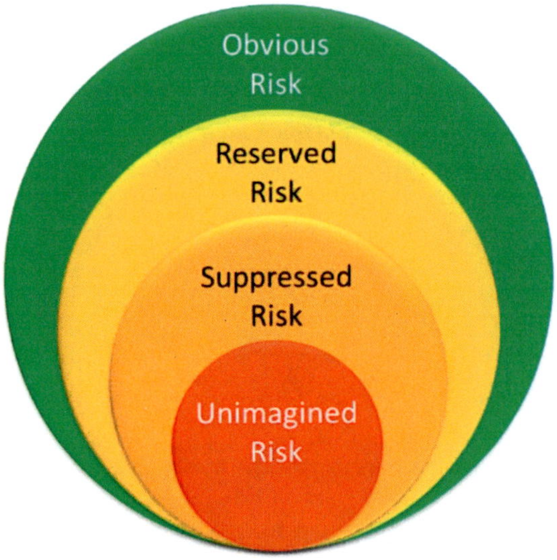

Obvious risks are risks that are readily apparent and easy to identify. These are the risks that your participants freely volunteer. They are the risks that people are comfortable contributing. They are uncertainties that are safe and uncontentious. These are the low-hanging fruit of risk identification.

An example of an obvious risk might be "a key project resource may fall ill resulting in delays that cause the schedule to slip." This level of risk is common to almost any project. It is safe to call it out, as it is obvious.

Reserved risks are less likely to be contributed. They are risks that your participants are fully conscious of, but they may be contentious risks. These are the risks that you will need to work to elicit from your participants. These are the risks that you will need to tease out by prompting your participants with the tools and techniques in this chapter.

Suppressed risks are more difficult to elicit. These are risks that your participants are either not consciously aware of or they are risks that need to be teased out through logical chain of reasoning. They are the risks buried somewhere in the subconscious of your participants. Sometimes we need to explore the causal chain to discover suppressed risks.

Unimagined risks are risks that your participants have never thought of. There could be areas of uncertainty that some of your participants have never considered. You will need to use the tools and techniques in this chapter to challenge your participants to consider all areas of uncertainty. The use of causal chains, checklists, and categories can be especially effective for uncovering unimagined risks.

A ***causal chain*** is the path of influence running from a problem's root cause to problem's symptoms. Each link in the chain represents something in the real world. At one end of the chain is the root cause. At the other end are the symptoms that the risk causes. The links between the two ends are the intermediate causes. A simple example of a causal chain can be seen in the Diagram 16 below.

Diagram 16 - Causal Chain Diagram[23]

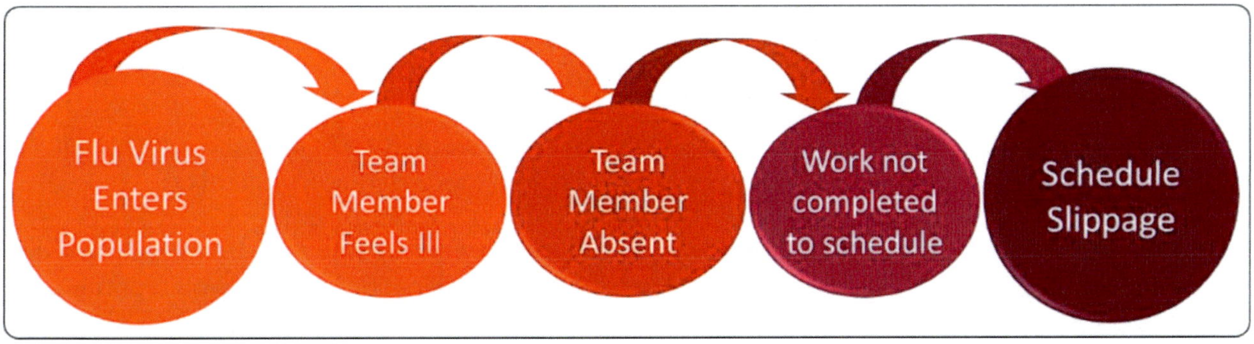

Use the tools and techniques in this chapter to help you tease out reserved, suppressed, and unimagined risks. You can use tools and techniques such as "the five whys" to prompt your users to consider the logical steps that will help them construct the causal chain needed to identify suppressed and unimagined risks.

For example, you may have a vendor that might seem stable and low risk, but they may be exposed to their own risks that could impact the risk that they pose to your project.

Consider the risk of a vendor who generates over half of their business from a single customer. Your vendor may seem stable and low risk when you interact with them, but what if their single large customer had an emergency and demanded that the vendor drop work on any other project to work on their project? How would the vendor react? They would probably drop work on your project to jump to the bidding of their dominant customer.

How would that impact your project? After we consider the causal chain, your stable and low risk vendor seems much riskier. Through the causal chain, your vendor's risk becomes your risk.

[23] https://www.thwink.org/sustain/glossary/CausalChain.htm

Or what if your vendor's single dominant customer went bankrupt and couldn't pay their outstanding invoices? How could your vendor's ensuing financial duress impact your project? Would your vendor's financial duress cause them to start cutting corners (i.e., reduce quality) on your project work to reduce their costs? Would your vendor go bankrupt, leaving your project in the lurch?

When facilitating the risk identification part of your workshop, the facilitator needs to use the tools and techniques in this book to elicit the reserved, suppressed, and unimagined risks.

If you simply let your participants drive your risk identification session, you will end up with a list of obvious risks. That would be of limited benefit to your project. Instead, the facilitator needs to coax, prompt, and cajole the participants to unearth the reserved, suppressed, and unimagined risks.

The obvious risks are the risks that would generally be dealt with in the normal course of managing a project. It is with the identification of reserved risks, suppressed risks, and unimagined risks that the risk identification exercise can really yield benefits.

Use the tools and techniques outlined in this chapter to ensure that you move beyond simply capturing the obvious risks. Use the tools and techniques outlined in this chapter to capture the reserved risks, the suppressed risks, and the unimagined risks

4.4.4 Setting the Context for Risk Identification

Axelos M_o_R tells us the goal of setting the context for risk identification is to get information about the activity that is planned and information on how it fits into the broader organization.[24]

By sharing information about your planned risk identification activities and how those activities fit into your workshop, you are priming your participations for risk identification. As you would prime a pump by adding liquid before beginning to operate it, you are adding information and structure to your workshop before beginning the risk identification.

The purpose of setting the context for risk identification is not to actually identify any risks. The purpose is to prepare your participants so that they can pump out a rapid flow of risks when the risk identification begins.

There are several tools and techniques that we can use to set the context of risk identification. These are

- Stakeholder analysis
- PESTLE analysis
- SWOT analysis
- Horizon scanning
- Lessons learned
- Pre-mortem

[24] Axelos M_o_R, pg. 32

We'll go through each of these in detail in the following sections.

4.4.4.1 Stakeholder Analysis

Regardless of where you are in your project lifecycle, you will likely have done at least some stakeholder analysis. This point in your workshop is an opportunity to both build on the stakeholder analysis you've already done and to get the group to review and refresh what you already know about your stakeholders.

The goal of stakeholder analysis is to capture information on the stakeholders' power, attitude, and interest. We want to identify both *how they could impact the project* as well as *if they would be inclined to impact the project*. This will inform the potential risks your stakeholders pose to your project.

If you have an existing stakeholder list, I would use that as a starting point. You could then use an org chart as a reference to check that you have identified all internal stakeholders.

When looking for external stakeholders, I like to use a simple affinity diagram to map out the project's external touchpoints. These external touchpoints help to identify one or multiple external stakeholders for each touchpoint. See an example of an external stakeholder affinity diagram under External Risk, section 4.5.7

Once you've identified all the stakeholders, it's a good time to run through each of them to determine their *power to influence*, their *attitude towards the project,* and *their interest in the project.*

Attitude: Are your stakeholders supportive? Are they detractors? Are they somewhere between? What do they WANT for your project?

Power (also known as influence): What power do they have to influence your project? Regardless of attitude, can your stakeholder influence your project for better or for worse? In other words, what can they DO with their attitude?

Interest: How much do your stakeholders care about your project? How deeply "invested" is each stakeholder in the project's outcome?[25] How much do your stakeholders CARE about your project?

[25] https://www.brighthubpm.com/project-planning/23481-stakeholder-analysis-spheres-of-influence/

Diagram 17 - Stakeholder Identification – Spheres of Influence Diagram

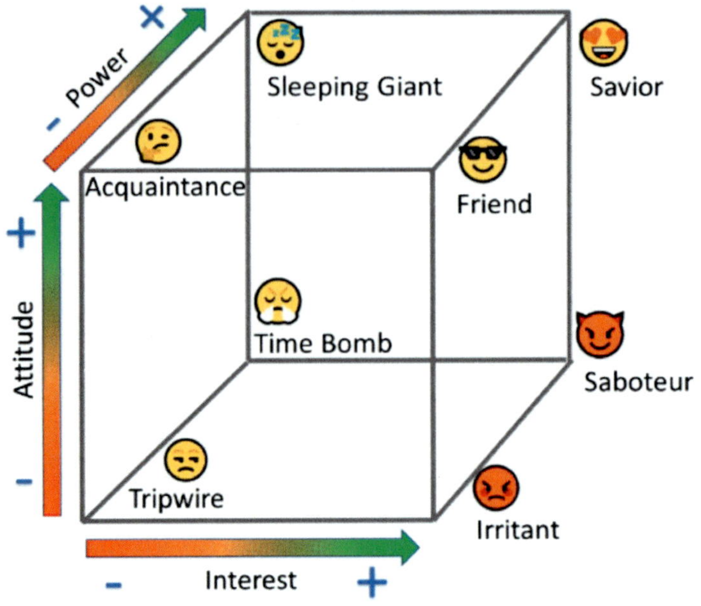

To inform any stakeholder related risks, stakeholders can be classified into categories according to their combination of attitude, power, and interest. These categories are described in the spheres of influence in the table below.

Table 6 - Spheres of Influence Table

Character	Icon	Power	Interest	Attitude
Trip Wire	😒	Low	Low	Negative
Time Bomb	😫	High	Low	Negative
Acquaintance	😕	Low	Low	Positive
Sleeping Giant	😴	High	Low	Positive
Irritant	😡	Low	High	Negative
Saboteur	👿	High	High	Negative
Friend	😎	Low	High	Positive
Savior	😍	High	High	Positive

Once you have your stakeholders categorized, keep the categorization visible, as they will help inform risk identification later in the workshop.

4.4.4.2 PESTLE Analysis

PESTLE analysis (Political, Economic, Social/Cultural, Technological, Legal, Environmental) is a classic prompt list that helps us to examine the project environment from a variety of external perspectives. PESTLE analysis helps you guide your participants to perform a wide scan of the project environment.

I like to walk through each element of PESTLE, asking for each element if there is any potential change coming up that could negatively impact the project. Then, I separately ask if there is any upcoming change that could positively impact the project.

Political

Are there upcoming elections or appointments that might affect your project — municipal, state/provincial or federal? Are there any changes to the executive, upper house, or lower house? Could these affect your project?

If your project spans national boundaries, are there any upcoming changes in any of the countries that your project operates? Are there any potential changes in how these countries relate to each other?

Are there any political shifts within the current political environment? Or any changes of leadership, focus, or emphasis that might affect your project?

Are there any other changes to political power dynamics that might impact your project? Are there any international political changes that could impact your project? Is your project dependent on any international suppliers who could be impacted by politics?

Economic

Are there any likely changes to the economy that could impact your project?

Where is the economy in the business cycle? What is your project's risk exposure to an economic downturn? What is the risk exposure to an economic upturn?

Are there any other factors that might have a big enough impact on the economy that might affect your project?

Social / Cultural

Are there any social trends that may impact your project? Are there any trends that could be a risk to the project? Is the group aware of all potential cultural norms that could constitute a risk?

Are there any upcoming changes in preferences that may impact your project?

Are there any cultural sensitivities that may affect your project? There are many examples of projects that have run into trouble due to social or cultural misunderstandings, especially when working internationally or when having international stakeholders.

Are there any potential cultural sources of conflict amongst your stakeholders? Some cultures hold cultural grievances against people of other cultures caused by incidents that happened decades or even centuries in the past.

Are there any cultural sensitivities that could impact your project? I recently spoke to a project manager in Saudi Arabia who was working on a large tourist development project on an undeveloped island. He said their project was subject to risks that stemmed from particular cultural sensitivities.

The project's poorly paid migrant construction workers from the Indian sub-continent had not been provided with adequate sanitation facilities. They were resorting to using the beach as their lavatory. As a result, the beach was littered with human excrement.

Apparently, the upper class of Saudi Arabia has an acute cultural revulsion to any exposure to human excrement.

The project risk register held a risk that a member of the royal family might come for a site visit and might see human excrement. This risk merited a high risk score. It is a result of the particular socio-cultural sensitivities that such a risk would be considered a high risk.

The above example was specific to a construction project, but similar considerations are possible in ICT projects. Examine your project's environment to see if there are any socio-cultural considerations that may pose risks.

Technological

Are there any technological considerations that could trip up your project? Technological considerations often have dependencies that must be understood to identify the risks.

Are there any lead times required for technological activities? Is there any preparation work that needs to be performed before technological activities can be started? Are there any lag times that the project would have to wait for after a technological activity is completed?

Are there any competing technologies that could cause issues for your project? Could your project be impacted by a disruptive technology?

Is the technology well understood, and do you have the resources to address technological issues if they arise?

Environmental

Could your project impact on the environment in any way?

Are there any environmental regulations that would restrict or impact your project's activities?

Are there any environmental requirements that would affect your project or any compliance activities that you need to plan for?

Could your project positively impact the environment in any way? Could your project negatively affect the environment in any way?

Legal

Are there any upcoming changes to legislation that could impact the project? Is there any legislation that you may run afoul of? Are there any regulations or laws that you may breach?

Is there any current or upcoming lawsuit that might affect your project? Does your project have any exposure to potential lawsuits? Does your vendor have any exposure to potential lawsuits?

Are there any lawsuits external to your project that could set precedents that could impact your project?

Are there any potential changes to the enforcement of laws or regulations that could impact your project? I've worked on a project where existing legislation that had not been enforced for years was suddenly actively enforced by the regulators. This sudden change in enforcement caused a schedule delay.

Ask if there are any regulations that are not being enforced, and also ask if there are any regulations that are being enforced that may not be enforced in the future. Changes to government department or regulator's budgets often impact on their ability and inclination to enforce regulations and laws.

4.4.4.3 SWOT Analysis

A SWOT (strengths, weaknesses, opportunities, and threats) analysis is a classic technique that juxtaposes internal characteristics against external factors. SWOT analysis helps focus your group on the strengths and weaknesses of your project and organization. It then allows you to probe those internal strengths and weaknesses to identify how they can interact with external opportunities and threats.

Analyzing strengths, weaknesses, opportunities, and threats together combines the results of both internal organization analyses and external environmental analyses.

As a result, a SWOT analysis is beneficial for both setting context as well as for actually identifying individual project risks. As you go through the exercise, make sure you record any risks that your participants call out.

I find the easiest way to use the tool is to collectively identify the strengths and weaknesses first. I like to use a whiteboard and to ask participants to identify the strengths and weaknesses for the project first. Then, I like to ask what organizational strengths and weaknesses the project could draw on to help achieve the project goals.

Diagram 18 - SWOT Diagram

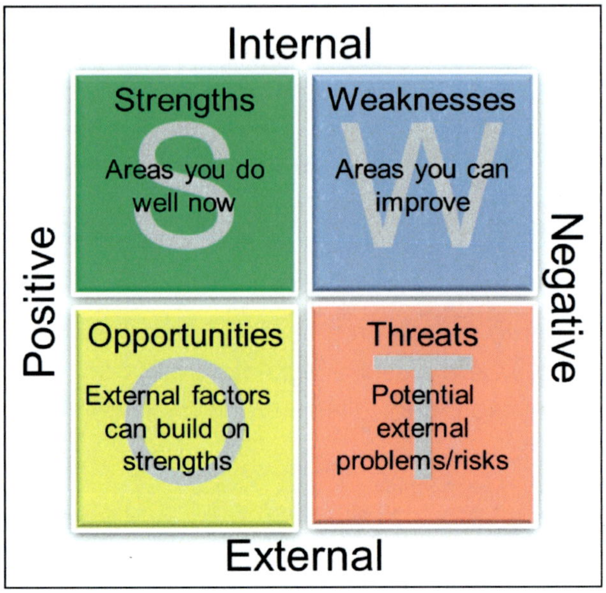

Once you have a list of strengths and weaknesses, I like to run through each of them to ask, "How does that pose a threat?" and "How does that provide an opportunity?"

The results of your SWOT analysis can be helpful as a reference for the rest of the workshop. I like to keep the results on a whiteboard so that everyone can reference it throughout the workshop.

4.4.4.4 Horizon Scanning

> *"Every person takes the limits of their own field of vision for the limits of the world."*
>
> Arthur Schopenhauer, German philosopher

Horizon scanning is simply examining likely future developments that are at the outer margins of current planning. Horizon scanning involves probing the edges of current thinking to identify potential areas of uncertainty.

I find the easiest way to start horizon scanning is by describing the project objectives and asking, "Is there anything on the horizon that could potentially impact our project?"

If participants identify any areas of uncertainty without any further prompting, then you're off to a flying start.

Once the voluntary flow of ideas dries up, I then run through categories (listed in Appendix A) as a way to probe specific aspects of the project to see if there are areas of uncertainty on the horizon in any particular category.

I find it works best to start with open-ended questions and see what your participants come up with. Ask questions like "Are there any financial events or changes on the horizon that could impact our project?"

Once they've exhausted their ideas with the open-ended prompts, you can then use more specific questions to further explore categories to guide and probe them for further ideas. You can use specific questions like "Are there potential changes to trade tariffs that could financially impact our project?"

For example, if you're considering the category of *communication*, you could ask specific questions like

- "Are there any current communication methods that may be falling out of favor?"
- "Are they trending down?"
- "Are they being replaced?"
- "Is the technology coming to end of life?"
- "Do these trends pose risks to the project objectives?"
- "Are there any new communication methods that might be coming into favor?"
- "Are they trending up?"
- "Are they replacing any existing technology?"
- "How do these trends provide opportunities for the project?"

4.4.4.5 Lessons Learned and Previous Risk Registers

Lessons learned and risk registers from former projects can be a great resource to draw from to prime your participants to consider uncertainty. Reviewing what went wrong in other projects can be an effective way to stimulate your participants to consider uncertainty more broadly.

I recommend curating the lessons learned and historic risks before you hold your workshop. This will allow you to eliminate material that is redundant or irrelevant. It ensures you only present material that is applicable to your participants.

4.4.4.6 Pre-mortem

The final tool I like to use when setting the context for risk identification is a light-hearted tool called a pre-mortem.

As a post-mortem is performed after death, a pre-mortem is performed before death. But instead of performing it on a human body, we perform the pre-mortem on our project.

In a pre-mortem, we imagine a future in which our project has failed, and then we work backwards to determine what potentially could have caused the failure. There are a few stages to a pre-mortem:

First, set the stage – Introduce the concept of the exercise. Encourage your participations to let their imaginations run wild, but within the constraints of the project. Choose a future date and describe the failed state of the project.

Second, determine how full your glass is – Remember that risk is about uncertainty, both positive and negative uncertainty. Separate your participants into two groups. One group is your "glass is half full" group. They're going to look for all the positive uncertainties that could happen with your project.

The other group is your "glass is half empty" group. They're going to look for all the negative uncertainty that could impact your project.

Third, gather ideas – If you have a whiteboard, encourage your participants to use their imaginations and jot everything down on sticky notes to put onto the whiteboard. This isn't the time to monitor the quality of responses; this is the time to encourage open thinking. The ideas contributed may spark thoughts on new undiscovered areas of uncertainty.

4.5 Identify the Individual Risks

Setting the context for risk identification should have primed your participants for risk identification. Now, while your participants are thinking broadly about risks, it's time to run them through the tools and techniques we use to pump out a stream of risks.

It is likely that your participants have already come up with some risks while going through the process of setting the context of identification. If so, you're off to a great start. You will be able to slot those risks into the identification categories as you go through the identification process.

4.5.1 Individual Brainstorming Incorporating Modified Delphi Technique and Nominal Group Technique

As mentioned earlier, collective brainstorming is a group effort that aims to generate output that is greater than what could be achieved by the total efforts of the individuals.

Conversely, individual brainstorming consists of inviting each participant to write down their ideas *before* seeking any spoken suggestions from the group. This may seem counter to the purpose of conducting a workshop, but please allow me to explain.

In section 4.1 we discussed various factors that can inhibit or influence an individual's participation. Regardless of the reason for a participant's disinclination to participate, it is your challenge as the facilitator to minimize any inhibitors and to get everyone in your workshop to contribute. Incorporating individual brainstorming into you're the outset of your collective brainstorming is a powerful strategy for overcoming many common challenges to stakeholder participation.

Consider *first-mover* bias, as described in section 4.1. First-mover bias can skew the output of your workshop towards the first idea suggested. Once a participant shares the first idea, other participants' thought processes skew towards the first topic suggested. The first idea can become an anchor for the rest of the workshop, an anchor that people keep coming back to.

For example, if a participant from the finance department starts the workshop by airing concerns about costs, it is likely that the group will suggest more cost-related risks and fewer risks related to other areas.

You can use individual brainstorming at the outset to lessen the impact of first-mover bias. Ask your participants to write down the uncertainties that are on the top of their minds <u>before</u> you invite anyone to share their thoughts with the rest of the group. This ensures that your participants' initial thoughts are not lost to the effects of first-mover bias.

This individual brainstorming also enables each individual to create their own anchor. If one individual thinks of cost uncertainties first, they will likely skew their brainstorming towards costs for the rest of the workshop. At an individual level, that is not as much of a problem, as other participants will create their own anchors on quality, scope, schedule, etc. In aggregate, you will still have a broad range of uncertainties that have been considered.

Individual brainstorming can also be used to good effect with participants who are reserved. Reserved individuals are much more likely to contribute when given the opportunity to contribute their thoughts anonymously.

If you organize your workshop so that your participants will not be questioned on their contributions, you will get broader and deeper contributions from your reserved participants. I recommend you manage brainstorming as described below.

Before seeking any spoken suggestions from the group, invite each participant to write down at least three areas of uncertainty on sticky notes—one idea per sticky note. Allow the participants five to ten minutes to jot down the risks that are front of mind.

When time is up, ask everyone to stand and then place their sticky notes on a whiteboard or a wall that all can see.

Before the workshop, I like to provide categories for people to stick their sticky note under. I find it makes the rest of the workshop easier to manage (see the example categories in Appendix A). If you'd like a ready-to-print set of *category headings*, you can pick them up at www.ictrisk.com.

Categories stimulate thought and provide a way to organize the individual risks. Before the workshop, I like to write the categories on the board or print them and stick them to a wall. Participants can then choose the best category to post each of their sticky notes under. This saves time in aggregating and categorizing and also helps to stimulate thought.

Once the wall or board is filled with risks and people have had a chance to mill around and look through them, it's time to open the floor to further contributions. Go through each category and read each risk for the group. Hearing the risks that their colleagues have contributed often sparks more suggestions.

As a facilitator, you want to encourage this free flow of ideas as this often has a compounding effect. The more risks that are suggested, the more ideas trigger additional risks to be identified.

Before moving from one category to the next, ask if there are any more risks related to that category. Ensure that everything is recorded, ideally on the same board or wall. You want to ensure that all participants can view the ideas that have been contributed and to see which categories may have been neglected.

After you have run through all the categories, you will have to decide a point to rein in the conversation. There is a fine balance between managing the workshop to your agenda and missing out on a natural flow of thought. Throughout the identification process, keep an eye on the clock to ensure you have enough time to run through everything on your agenda, and ensure that all contributions are being captured by your scribe or note-taker!

While risks are being discussed, take the opportunity to ask questions of risks that seem too high or too low a level. Don't get bogged down in the detail, but it is worth asking a quick, polite question for risks that are obviously at the wrong level.

For example, if "risk of failure to deliver to the schedule" is suggested, you could ask which tasks on the critical path might fail and cause that schedule delay. People usually know the underlying issues, but they may have difficulty in articulating the uncertainty. Questions are the keys that help them unlock the answers.

The purpose of querying the level of risk is to draw reflections from participants. The purpose is to encourage them to query for themselves whether risks are at a level such that meaningful action could be assigned to them. If a risk is set at a level that is too high, encourage participants to break it down into its component parts. Always put the questions tactfully and avoid saying anything that could be interpreted as criticism.

Similarly, if a participant suggests a risk that is very granular, ask if it can be rolled up with other risks that could be treated by a common risk response plan. For example, if the availability of a single person is suggested as a risk, you could ask if it could be rolled up to the availability of a team (provided the risk response plan would be the same for the whole team).

Projects and organizations vary too much to be prescriptive on this point. It up to you to provide your participants with gentle guidance if the risks suggested are clearly too high or too low.

Probing

Once the flow of ideas starts to dry up, but people are still in the brainstorming mindset, it is time to coax, prompt and cajole the participants to see if there is any uncertainty that has been left out.

Hopefully, you now have many risks spanning the majority of the categories you've used. Your participants have racked their brains and have come up with as many areas of uncertainty as they could. Now is your opportunity to add value by using the tools and techniques in this chapter to stimulate thought and see if you can unearth risks that have not yet been considered.

People's thought processes vary from individual to individual. The tools and techniques in this book take disparate approaches to probing and cajoling people to think about uncertainty. These tools and techniques overlap, but they are complementary. They address the same questions from different angles. At least one of these tools and techniques should stimulate thought in each of your participants.

All the tools and techniques approach uncertainty from different angles or perspectives. As such, there will be some overlap. This overlap in itself may stimulate thought as the same areas may be considered from different perspectives.

Avoid Rushing

One danger to watch out for in your workshop is people wanting to rush through risk identification if they feel they don't have enough time. When people are under time pressure and stressed, their vision tends to narrow. Questions like, "What are the key things we must do?" can lead to narrow thinking that can seriously reduce the value of your risk identification efforts.

If you run short of time, don't fall into the mistake of succumbing to catch phrases like "80:20 rule," "focus," "prioritize," and "critical success factors." Such catch phrases narrow thinking and reduce the effectiveness of risk identification.

Instead, manage your time throughout the workshop to forestall any sense of urgency and encourage broad, open mindsets.

You may have to run through all the easy and obvious risks before people start to identify the edges of uncertainty. Allow enough time to cover the easy and obvious first before probing for the more difficult areas of uncertainty.

The real danger often lurks within the difficult-to-identify 20% of the project risks. The easily identified 80% are more likely to be handled within the project's normal course. In risk management, it is not the low-hanging fruits that cause trouble; it is the hard-to-reach 20% nestled in the upper branches where the real danger lurks! Keep to the agenda and work your way through the probing tools to ensure you capture as many risks as possible.

In the following section, we explore the tools and techniques that you can use to prompt participants to identify more risks. I recommend holding back these prompts until after the natural flow of new ideas from brainstorming has dried up, lest the prompts influence and divert your participants' natural flow of thought.

Once the natural flow of ideas has run out, tools and techniques provide an effective prompt to uncover the more difficult to identify risks. They encourage participants to identify risks that hadn't been considered before. Keep in mind that the objective is to stimulate thought about uncertainty. Even if one aspect of a tool or technique is not directly related to your project, it might inspire a participant to think of one that is.

4.5.2 Root Cause Analysis

Root cause analysis (RCA) is a tool that helps us to explore different potential causes and impacts. Causes and impacts may help to flesh out the risk description at the most appropriate level. And RCA may help us identify more risks. This can be especially effective in interrogating any risks that look like they may be at the wrong level.

The *risks* that participants identify during individual brainstorming are sometimes actually *causes of risks* or *symptoms of risks*. In these situations, you can ask the appropriate questions to get to the actual risk.

If a participant suggests a cause, you could ask, "What could that lead to?" Or, if they suggest a symptom (effect), you could ask, "What could have caused that?" It is your job as facilitator to capture the correct portions of the causal chain without criticizing the input or the participants.

If someone suggests a risk that the project delivers late, you could ask what would cause the project to deliver late. The project delivering late is actually an effect of a risk. There are likely quite a few different risks that could cause the project to deliver late.

For example, there is a risk that the duration required to complete a critical path task may have been underestimated. Or, there is a risk that key resources are not available when they are needed. Or, there is a risk that a supplier is late on delivery. Each of those risks could impact the project by causing it to deliver late. The root cause analysis technique can help you uncover the underlying risks.

RCA can be used to identify threats and/or opportunities. You can do this by starting with a problem statement (for example, the project might be delayed or over budget) and then explore which threats might result in that problem occurring. The same technique can be used to find opportunities by starting with a benefit.

Root cause analysis (RCA) can be addressed using many techniques.

4.5.2.1 5-*Whys* Analysis

One of the techniques I like to use in RCA is the *5-whys analysis*. The 5-whys analysis is a simple problem-solving technique that helps users get to the root of the problem quickly. It was made popular in the 1970s by the Toyota Production System.[26]

The 5-whys technique involves examining a problem and asking "why?" and "what caused this problem?" Often the answer to the first "why" prompts a second "why." The answer to the second "why" often prompts a third "why" and so on. When you ask "why" to each answer, you drive your participants to think more deeply about the root cause of the uncertainty. Questions are the keys that help your participants unlock the answers.

The 5-whys analysis is a simple and quick method to guide your participants to identify what the root cause is of each of the risks that they identify. I don't use the 5-whys analysis on every risk suggested, but I do engage the technique when a risk is suggested that sounds like it's not describing the underlying risk correctly. The 5-whys analysis is one of the most effective techniques for unpacking risks to discover the underlying root cause.

Determining when to use the technique is subjective, but you'll soon find when it's useful and when it's not. If in doubt, it's better to use the technique and waste a couple of minutes than to not use it and let an important cause go undiscovered.

Diagram 19 - RCA Using 5-Whys Analysis Diagram

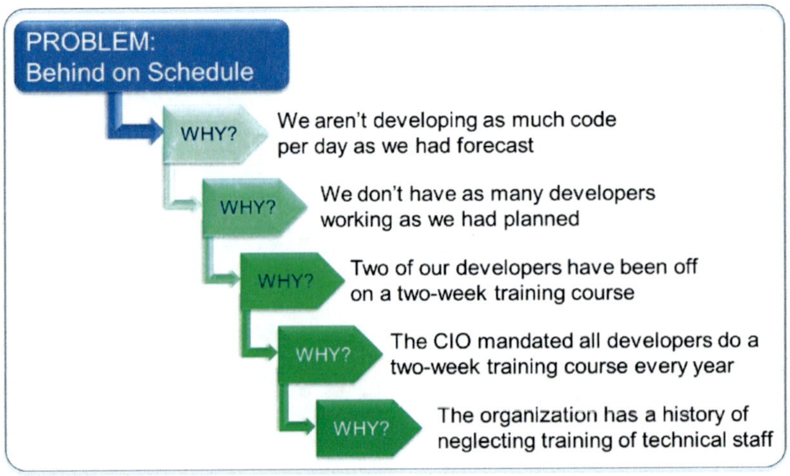

[26] https://en.wikipedia.org/wiki/Five_whys

4.5.3 Cause and Effect Analysis (also known as Ishikawa or Fishbone Diagrams)

A cause and effect analysis helps to develop an understanding of causes or sources of uncertainty that may give rise to risks. It helps identify individual areas of uncertainty that have many branches of causality.

Some people use the same tool to first identify the causes and then work from the causes to the results. That is perfectly fine too. The use of the tool is not prescriptive; you can use it in whatever way works for you.

But the way I have found it most effective is to work backwards from the consequence to the causes. I usually use one diagram for each of the project objectives (e.g., cost, schedule, quality, and scope). Then I use the area above the spine to identify negative causes of uncertainty and the area below the spine to identify positive causes. This is a format I have found effective. But, feel free to experiment and find your own way to use the tool.

Often participants will start by identifying high level elements in the causal chain (see example in section 4.4.3). The Ishikawa tool is designed to cope with that.

You can list high-level elements closer to the spine. Then, you can borrow from the 5-whys technique (see section 4.5.2) to question how those high-level elements decompose into elements further in the causal chain. These detailed elements can be recorded on the diagram as bones that stem out further from the high-level bones (as per diagram 20 below).

Diagram 20 - Ishikawa Diagram

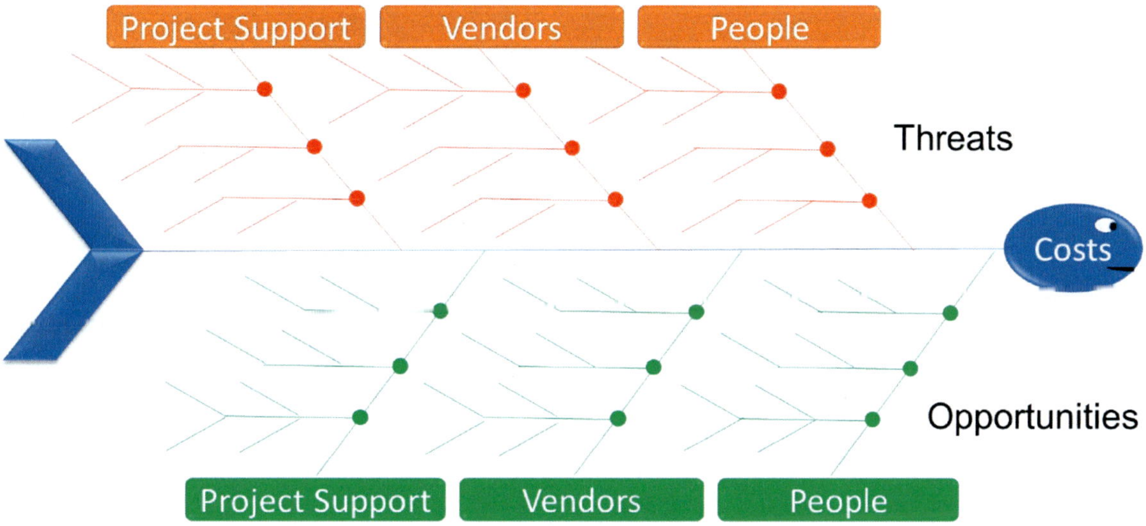

4.5.4 Project Management Constraints

A review of the project constraints categories may stimulate thoughts that generate awareness of more risks. Run through each of the constraint categories in turn, reviewing their current state. Ask if there are any more risks in each constraint.

There is no consensus about what constitutes a category of project management constraint, so feel free to modify the list below. Remember that the purpose of this exercise is to probe thinking about undiscovered uncertainty. It is far better to have too many constraints in your list than to have too few. I suggest starting by checking for constraints with the following six categories as described in section 1.3 Project Constraints:

- Schedule
- Budget
- Quality
- Scope
- Resources
- Benefits

4.5.5 Assumptions Analysis

If we think about it, every assumption contains a risk. We just need to tweak the wording and syntax to draw that risk out so that it doesn't trip us up. Remember the old adage about the dangers of "assume", as per the diagram 21 below.

Diagram 21 – The dangers of assumptions

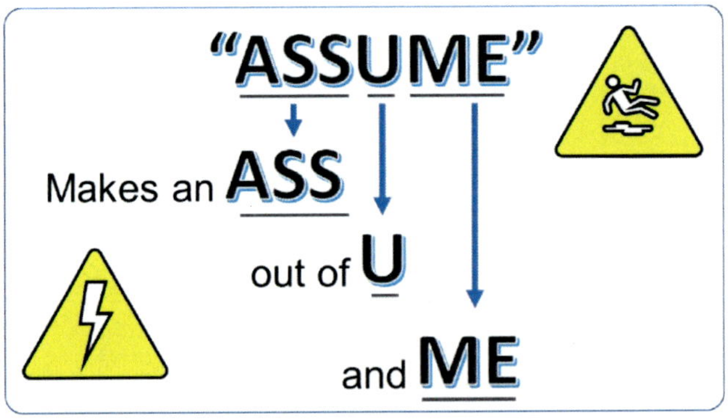

For example, if your project A has a project assumption that you will be able to use the shopping cart functionality that project B is creating, you can identify the risk hidden in that assumption by asking "What if that assumption turns out to be wrong?"

Your risk might be "Project B builds a shopping cart that does not meet our project requirements, resulting in delays that cause our project schedule to slip."

In your workshop, run through each of your project assumptions. Ask your participants "What is the risk if this assumption turns out to be incorrect?" or "What risk is embedded in this assumption?"

4.5.6 Project Constraints Analysis

Project Constraints differ from the *Project Management Constraints* discussed in section 4.5.4. Project Constraints are limitations associated with a project that either must happen or must not happen. As with assumptions, constraints often contain risks. As such, constraints analysis is very similar to assumptions analysis.

Each constraint begs the question, "What impact could this constraint have on achieving your project objectives?" Therein lies the risk.

Once your participants have supplied you with an answer, it is up to you to reword and adjust the syntax to describe the risk embedded in the constraint.

For example, you might have a constraint such as "The project will have to operate within organizational release management processes."

The risk embedded in such a constraint might be reworded as "The mandated use of the organizational release management processes could cause delay in releasing the product into production, resulting in delays to the project schedule."

4.5.7 External Risks

In risk identification, stakeholders tend to concentrate on internal risks. Calling out the differentiation between internal and external risks can help stimulate your participants to think about the external risks.

Look for any touchpoints your project has with the external world. Ask if there are any uncertainties for each touchpoint to cover all possible interfaces. If you have a diagram showing your project's external touchpoints, it can be a good visual when trying to identify external risks and their interest in your project. Consider

- Customers
- Vendors
- Contractors
- Consultants
- Regulators
- Lawyers
- Industry bodies
- Unions
- Users
- Media
- All levels of government

Each one of these external touchpoints may have several subcategories to consider. See the affinity diagram excerpt below for an example of a small portion of an external risk assessment. Unfortunately, a fully completed diagram would not fit on a page. In the example in diagram 22, *project* serves as the central hub node. The project would serve as a hub on a wagon wheel, with other nodes spoking off in a circle encompassing the hub, as *media*, *gov't* and *users* do in the example below.

Each of the nodes in the first level around the Project hub may have sub-components. For example, Gov't (Government) may split off into Municipal, State and Federal.

Diagram 22 - External Stakeholders Affinity Diagram

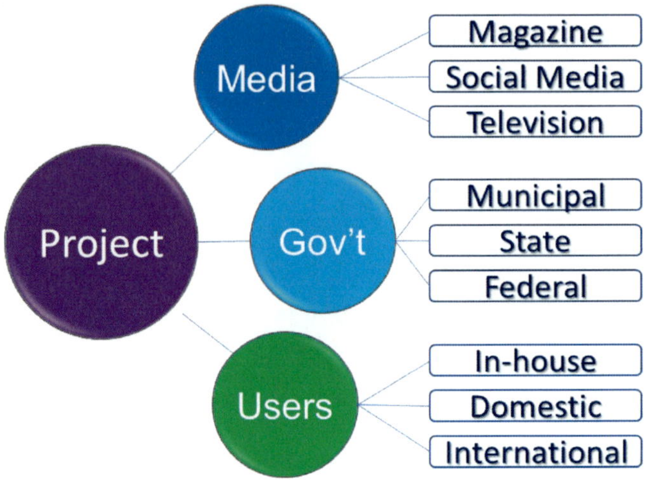

4.5.8 Mind Maps vs. Affinity Diagrams

Mind maps and affinity diagrams are simple tools that help us to think about a logical structure to the relationships between ideas. They are both versatile tools that can be used in many ways.

Affinity Diagrams

With affinity diagrams, ideas are logically organized based on their relationship to each other. Affinity diagrams work by taking series of facts, data, and ideas under a particular theme and organizing them into different clusters based on their affinity or relationship with one another.

For example, if I'm working on a software development project, I may organize my tasks based on the things I need to create. I may organize in a sequential order tasks for registering new applications, modifying existing applications approving applications, printing, reporting, etc.

Mind Maps

In contrast, a mind map is less logically structured. While affinity diagrams have a logical organization, mind maps capture ideas as they come up, and those ideas may or may not relate to each other. Mind maps make it easier to visualize data or discover relationships between ideas.

Which diagramming technique should I use?

I find that mind maps are more useful when the subject is conceptual and immature. Mind maps provide the flexibility to explore a subject.

I prefer affinity diagrams when concepts are more mature, i.e., when concepts have been analyzed. Affinity diagrams also make it easier to manage concepts for a large complex project where mind maps may become too cluttered to be useful.

I find mind maps to be more useful in developing ideas and affinity diagrams more useful in communicating ideas.

I find that mind maps are more appropriate at the early stages of a project, and affinity diagrams are more useful later in a project. But, these are very flexible tools. Feel free to use which ever diagramming technique you deem most appropriate for your situation.

See below for a partially complete mind map. As you can see, mind maps can quickly become overcrowded and difficult to follow. It's best to find the largest surface you can when you create a mind map in your workshop.

Diagram 23 - Mind Map Diagram of Areas of Project Uncertainty

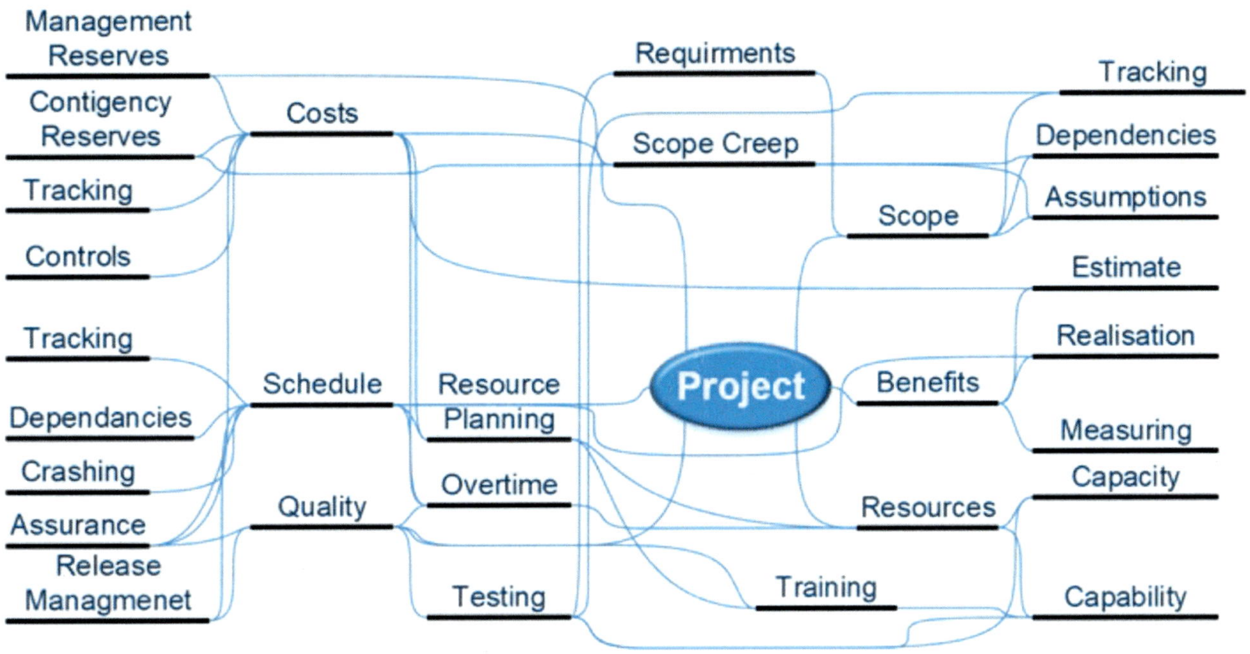

4.5.9 Checklists

Checklists are useful aids that can funnel organizational and industry learnings to prompt your participants to identify risks. They provide a structured method to ensure that mistakes made in previous projects become learned lessons in current projects. Checklists help to ensure previous project mistakes are not repeated in your project.

Checklists are the least flexible of all the tools and techniques discussed in this chapter. I recommend using them at the end of your risk identification session. Their rigidity constrains thinking, but they are useful at the end of risk identification to check that you've not missed anything.

If you have an organization or industry checklist, you could run through it with your participants to see if it sparks them to identify additional areas of uncertainty. If you don't have a checklist, feel free to use the checklist I provide in Appendix B.

4.5.10 Business Processes

If relevant and available, a review of the organization's high-level processes may be helpful when uncovering risks. If your organization has a process classification framework (PCF), it will provide an ideal visual tool to stimulate thought in your participants.

You could query each node in your PCF as you work your way through the framework to see if there is any uncertainty that could constitute a risk to your project.

4.6 Refine Risks

Once you have completed the risk identification exercise, it is good practice to take the time to refine your risks.

During the identify risk phase, you'll have recorded anything that your participants have suggested. You have tried to find the risks in the causes and effects that have been suggested as risks. You'll have attempted to catch any duplicate risks. You'll have aimed to articulate all risks to the most appropriate level.

All this will save you time in the refine risk process. But, it's almost guaranteed that during the fast-paced risk identification session, the risks you've captured will be a bit rough. It's almost certain that some risks will need to be refined before they can be effectively analyzed.

The refine risk process requires a different mindset from risk identification. To get the most out of your participants, you need to help them shift out of their blue-sky brainstorming mindset and into a more critical and analytical mindset. That's timely, as the following process—the qualitative analysis process—will also need participants to be in an analytical mindset.

A critical and analytical mindset will help your participants examine the output generated to date and to polish it in preparation for the qualitative analysis process.

4.6.1 Refine Risks Process

When refining risks, I like to work my way through each category of risks with the following steps:

1. De-duplicate - Identify any duplication and amalgamate where required
2. Question cause/effect - Question anything that sounds like a cause or effect to find out what the relevant risk is
3. Apply common syntax - Edit the risks to a common syntax
4. Roll up low-level risks - Amalgamate risks that are at too low a level (amalgamate with those that have the same treatment)
5. Break down high-level risks - Decompose risks that are at too high a level
6. Recategorize - Move risks from one category to another as required
7. Update register - Update the risk register (if there is a separate workshop for analyzing risks and developing the risk response plans).

4.6.1.1 Identify Duplication

Some of the risks that will have been identified are likely to be duplicates that have been articulated differently. You can ask for more detail on any risks that look like they might be duplicates to see if they are the same risk.

Any partial duplicates can be combined into new wording that covers each of the original risks. Any complete duplicates can be deleted if they are already covered by the wording of the other risk.

4.6.1.2 Apply Common Syntax / Metalanguage

Despite providing examples of syntax at the outset, you will almost certainly have some risks whose syntax needs to be rewritten. Take this opportunity to refine the risk syntax.

Using a common risk syntax will make the whole risk management process easier. Try to use the *cause-event-effect* syntax recommended by both PMI and Axelos (see description in 4.3.9).

4.6.1.3 Roll Up Low-level Risks

There may be some risks that are written at too low a level (see section 4.3.9). These risks need to be rolled up into a more appropriate level. Look for suggested risks that could be combined if they would have common risk response plans if they were to be mitigated.

4.6.1.4 Break Down High-level Risks

There may be some risks that are at too high a level to be useful (see section 4.3.9). High-level risks act like an umbrella, shielding and obfuscating one or many risks below them.

Look for risks for which you could not practically create a risk response plan for. Break them down into two or more risks, defined at a level where you could devise a meaningful risk response plan for them.

I find the “5-whys” analysis tool discussed earlier in Root Cause Analysis (see section 4.5.2) is effective in breaking down risks that are identified at too high of a level. The 5-whys analysis gives you a Socratic method of probing your participants until they uncover the risks that hide under the high-level umbrella risk statement originally identified.

4.6.1.5 Recategorize

You will likely have some risks that have been miscategorized. Take this opportunity to review the risks under each category. Move into the most appropriate category any risks that have been miscategorized.

You will likely have some risks that span several categories. Check with your participants and put these into the most appropriate category.

4.6.1.6 Update the Risk Register

After you've been through the refine process, your lists of risks should be much more manageable. They should have a common look and feel and will be much easier to work with in the analysis processes.

Your lists of risks should be in suitable shape to be added to the risk register, albeit in draft status.

If you're running separate workshops for the evaluation of risks, then it is a good practice to write the risks up as soon as possible after the identification workshop. You can then circulate them to the participants of the workshop and the broader community of stakeholders for comment and approval before changing the risk register status from "draft" to "open."

4.7 Chapter Summary – Risk Identification

There are challenges to risk identification. You will need to be aware of these challenges to devise the most appropriate strategy to overcome them. These challenges include

1. Reserved participants - For any number of reasons, you will likely have participants who are disinclined to suggest any risks. You will need to devise a strategy to draw them into participating.

2. Mindset bias – There are some different factors that will lead to mindset bias in your participants. You will need to devise a strategy to offset mindset bias to identify the broadest range of individual project risks.

3. First mover bias – You will have to develop a strategy to offset the natural inclination to emphasize the first subject that is suggested in your workshop.

4. Hijackers – Inevitably, in a large group of stakeholders, people will start to discuss things that are not relevant to the purported purpose of your workshop. You will need to devise a strategy to both minimize these occurrences and to quash them when they arise.

Preparation prior to execution is the key to facilitating an effective workshop. In this chapter, we have discussed preparing in the following areas:

1. Participant selection – You need enough people in the workshop to cover all aspects of the project, but you also need to limit the number so that the group is manageable.

2. Invitation – Getting a large group like this together is always a challenge. Work the schedule and get your invitation out early.

3. Agenda – You are going to manage a large group of people through a lot of material. Create a time-bound agenda and ask your participants to agree to stick to it.

4. Pre-reading – Prime your participants early so they're ready to go when the workshop starts. Give them enough information to cover the project but not so much that they are overwhelmed.

5. Presentation – Use a slide pack to speak to during the workshop. This can provide visual aids that will help you set the context of the project, the context of the workshop, and inform on risk management in general.

6. Material – You'll need some physical material if you're going to use the tools and techniques in this book.

7. Facilities – Select a space that is roomy enough for your participants to spread out and has all the equipment you'll need to use the tools and techniques in this book.

Running the workshop itself is most effective when you prepare your participants to identify risks. You can prepare your participants for risk identification by running through

1. The context of the project – include the project overview, key project goals, project objectives, project deliverables, project milestones, project assumptions, project dependencies, project constraints, and high-level project schedule

2. The context of the workshop – including the purpose of the workshop, the aim of the workshop, the outcome of the workshop, and the outputs of the workshop

3. Introduce the workshop methodology – familiarize the participants with how the workshop will be run and set their expectations

4. Establish the workshop rules – gathering a large group of people together is always a challenge; take the opportunity to establish the ground rules and ask participants to agree to them upfront to minimize disruptions during the workshop

5. Introduce meeting roles and expectations – defining the roles and expectations at the outset of the workshop reduces ambiguity and helps to manage your participants' expectations

6. Describe the risk management methodology – describe the steps involved in the risk management process so your participants have clear expectations

7. Introduce the agenda – introduce the sequence of steps that you will take through the workshop, the time allotted to each step and the schedule of breaks to manage your participants' expectations

8. Define the terms – define the terms that will be used in the workshop including the meaning of words, the syntax of risks, the risk levels, the types of risks worth capturing, the risks that are already covered, and external broad-reaching risks

It is important to inform your participants at the outset of the risk identification methods that you will use. It is important to explain how interviews and questionnaires have been used and how their output feeds into the workshop.

Then, it is important to set the context for risk identification to prime your participants to identify the risks. The tools and techniques we use to set the context include

1. Stakeholder analysis – A review of the internal and external stakeholders of the project; a chance to build on any previous stakeholder analysis and to map your stakeholders to their individual spheres of influence

2. PESTLE analysis – An evaluation of the project's external environment examining factors such as political, economic, socio-cultural, technical, legal, and environmental

3. SWOT analysis – An evaluation of internal project/organizational characteristics (strengths and weaknesses) juxtaposed against external characteristics (opportunities and threats)

4. Horizon scanning - examining likely future developments that are at the margins of current planning; horizon scanning probes the edges of current thinking to identify potential areas of uncertainty

5. Lessons learned and previous risk registers – Reviewing what went wrong in other projects can be an effective way to stimulate your participants to consider uncertainty more broadly

6. Pre-mortems – A chance to imagine a future in which the project has failed, and then work backwards to determine what could have caused the failure; a pre-mortem includes setting the stage, splitting the optimists from the pessimists, and gathering ideas

The actual identification of individual project risks draws on a whole other set of tools and techniques. Those tools and techniques include

1. Individual brainstorming – When individual brainstorming is combined with modified Delphi technique and Nominal Group technique, it becomes a powerful offset to several of the challenges to risk identification

2. Root cause analysis – RCA using the 5-whys analysis allows the facilitator to drill down into high-level risks, using a cascading sequence of questions to uncover the root cause of the uncertainty

3. Project management constraints – Reviewing the categories of project management constraints with your participants can stimulate the identification of additional individual project risks; constraints include but are not limited to cost, schedule, scope, quality, resources, and benefits

4. Assumptions analysis – Hidden in every project assumption is a risk; we need to tease out that risk for each assumption; what would happen to the project objectives if each assumption was proven wrong?

5. Project constraint analysis – Like assumptions analysis, there is usually a risk embedded in each project constraint. We need to discuss each constraint with our participants to identify the underlying individual project risk

6. External risks – Participants often tend to focus on internal risks; calling out the project's touchpoint with the external world can be an effective way to identify individual project risks that relate to the external environment
7. Mind mapping and affinity diagrams – Are effective tools and techniques to draw out more risks; these are very flexible tools that can be used in many different ways
8. Checklists – You may have a checklist that is specific to your organization or industry that is useful at the end of risk identification to ensure that you haven't missed anything
9. Business processes – If your organization has documented businesses that can be rolled up to a higher level (e.g., a process classification framework) then it can be a useful tool to run through with your participants to see if there are any process specific risks

The final step in risk identification is refining the risks. Refining risks polishes the risks that you have already identified to make the analysis of the risks easier. Despite your best efforts, the risks you have identified will almost certainly need some refinement.

1. Identify duplication – Inevitably, you will have some level of duplication. You might detect direct duplication, or you might find that some risks overlap with other risks. This step is about deleting duplicates or redefining risks that overlap.
2. Question cause/effect – Question anything that sounds like a cause or effect to find out what the actual risk is.
3. Apply common syntax – The preferred syntax of risk should be in the form of "cause-event-effect." Examine the risks you have identified to see if you can amend them to fit into that syntax.
4. Rollup low-level risks – Some risks will have been identified at too low a level. If you identify a low-level risk, look to see if there are other similar risks that you can roll it in together with.
5. Break down high-level risks – You are likely to have some risks that have been identified at too high a level for a meaningful risk response plan to be applied to the risk. Look to break these risks out into their component parts. Consider using the 5-whys tool to deconstruct high-level risks.
6. Recategorize – You will likely have some risks that have been clearly put under the wrong category. But, you will likely have more risks that span several categories. Work through these risks with your participants to determine which category is the most appropriate.
7. Update the risk register – If you are running a separate workshop for the evaluation of risks and the determination of risk response plans, then you would update the risk register with the draft risks and circulate it for review.

5 Qualitative Analysis

Perform qualitative analysis is the third link in our risk management process chain of steps.

Diagram 24 – Qualitative Analysis

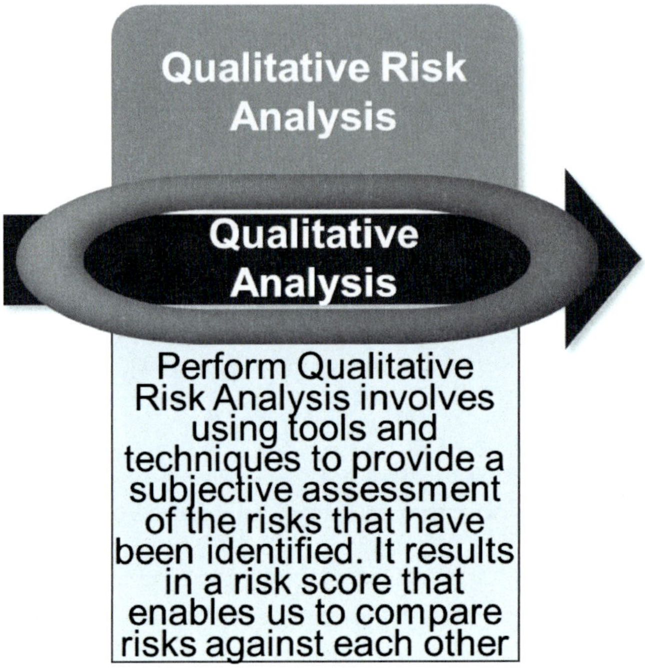

Now that you've identified and refined the risks, the next step in the process is to analyze the risks. You may want to do this in a separate session from risk identification to give participants a break and allow them to shift their mindset.

A break also gives them a valuable opportunity to mull over the risks. But, if you are time constrained, you may have to do risk identification, refine risk, analyze risk, and develop risk response plans in the same workshop.

The purpose of the *analyze risk* portion of the risk management process is to evaluate individual project risks both against each other and against predetermined metrics. Such metrics allow you to measure the individual risks against the organization's risk appetite and risk tolerance.

Qualitative analysis is the procedure of prioritizing individual risks for additional analysis or further action. We do this by evaluating their probability of the risk occurring and the impact that they would have on the project objectives if they did occur. Qualitative analysis may include other optional characteristics.[27]

The qualitative analysis process is the first stage of evaluating the risks that you've identified and refined. At a minimum, the process involves assessing the probability, the impact, and the priority of the risks.

Qualitative risk analysis informs the priorities of individual risks, and it is those priorities that determine how much rigor you put into the plan risk responses process.

Qualitative risk analysis also enables initial informed reporting. The output from qualitative risk analysis is the first time that you can give your stakeholders a meaningful view of the risks that the project faces.

Qualitative risk analysis is subjective, but it lays the foundation for quantitative risk analysis, which is objective.

The qualitative risk analysis process is not without its challenges. As qualitative risk analysis is subjective, you'll have to combat the subjective influences of groupthink and optimism bias throughout the process.

The primary benefit of qualitative risk analysis is that it enables you to focus on high-priority individual project risks. To do this, we need to be able to establish a method to score risks to determine which risks are high priority. And, to determine the risk score, we need to assess the probability of the event occurring and the impact that event would have on project objectives if it did occur.

Once we have established the probability and the impact for an individual risk, we can multiply the two of them together to get a risk score. This risk score allows us to compare disparate risks against each other. It enables us to determine the priority that will inform which risks we focus on.

The risk score enables us to determine which risks require further analysis, which risks require active risk response plans, and which risks go into the "watch list."

Perform qualitative risk analysis forms the comparative priorities of individual project risks in preparation for plan risk response.[28] This means that the output of the qualitative risk analysis will determine the level of rigor that will be applied to each risk in the plan risk response process.

We will examine risks that come out of the qualitative risk assessment process with a high risk score in more detail than those that come out with a low risk score. We will give these high score risks more consideration in the plan risk response process than risks with a low risk score.

But, that's not all the qualitative risk analysis does. The perform qualitative risk analysis process also lays the groundwork for the quantitative risk analysis process, if appropriate for your project.[29] The output of the qualitative risk analysis provides the input into the quantitative risk analysis.

[27] PMBOK Section 3.20 pg. 588
[28] PMBOK Section 11.3 pg. 421
[29] PMBOK Section 11.3 pg. 421

5.1 Risk Data Quality Analysis

Qualitative risk assessment includes risk data quality analysis for individual risks. Risk data quality assessment evaluates the quality of the risk data to determine if it is accurate and reliable. If the data is not of high enough quality to perform qualitative risk analysis, then it may be necessary to gather additional or better information.

Individual risk data should be assessed for objectivity, timeliness, and relevance. You can assess the quality of the data of any risks that seems questionable.

I find that it is easiest to use a questionnaire to assess the quality of a risk's data. You can ask your participants to fill in the questionnaire during your workshop and then calculate the averages afterwards to come up with data quality risk scores.

You can then use these data quality scores to determine which risks need further analysis. Risks that need further analysis can be added to the watch list as part of the risk register until more or better risk data can be obtained.

5.2 Additional Parameters

According to PMBOK, other parameters that you might consider including in your qualitative risk assessment are

- Urgency – How soon a response to the risk needs to be implemented if it is going to be effective; sooner would indicate more urgency
- Proximity – The amount of time before the risk might impact on any of the project objectives; a shorter amount of time signifies high proximity
- Dormancy – The amount of time that might pass after a risk has occurred before its impact is noticed; a short amount of time signifies low dormancy
- Manageability – The ease of managing the occurrence or impact of a risk; when management is painless, the manageability value is high
- Controllability – The amount of control a risk owner has over the risk's outcome; when the outcome can be controlled with little effort, controllability is high
- Detectability – The ease of detecting the impact of the risk being recognized; when the impact of the risk can be detected with little effort, detectability is high
- Connectivity – The degree that the risk is related to other individual project risks; when a risk is highly correlated or connected to many other risks, connectivity is high
- Strategic impact – The possibility for the risk to have an influence on the organization's strategic goals; where the risk has a significant effect on strategic goals, strategic impact is high

- Propinquity – The degree to which a risk matters to any stakeholders; when a risk is perceived as very important, propinquity is high[30]

All projects are unique. You may find some of these additional parameters useful and you may find some parameters less useful. Assess these additional parameters and determine which ones would benefit your project's risk management.

5.3 Risk Score

One of the first things we do in qualitative analysis is to determine the risk impact and the risk probability in order to establish a *risk score*.

Risk probability evaluation considers the likelihood that an individual risk will occur. Risk impact evaluation considers the potential effect the risk would have on any of the project objectives.

The risk score is determined by multiplying the impact value by the probability value of the risk.

The benefit of a risk score is that it allows you to compare two risks with very different characteristics. It allows you to compare a risk that has a high impact but low probability with a risk that has a low impact but high probability. A risk score provides you with a method of evaluating risks with very dissimilar characteristics against each other. A risk score enables you to compare apples to oranges.

Once you have established a score for each individual risk, you can rank the scores and determine their priorities. The aim of ranking risks is to inform decisions on what the risk responses plan should be, which risks should be addressed first, and how much effort should be applied to each risk.

The key benefit of qualitative risk analysis is that it allows your project to focus on the high priority risks.

Projects should pay more attention to, and dedicate more focus on, risks with a high risk score. You should examine high score risks in more detail. More time and effort will be put into determining their risk response plans for high score risks.

We also enact risk response plans earlier for risks with a higher risk score. Risks with a higher risk score are usually treated with more urgency than risks with a low risk score, and the risk response plans reflect that urgency.

For example, you may decide that a risk with a score of 16 or above must have a risk response plan enacted within a week. And, you may decide a risk with a score of 8 or above must have a risk response plan established within a month.

At the initial qualitative risk analysis stage, we are only looking at the inherent risk. But we should be clear at the outset on the different types of risk scores. To take a simplistic approach, for most projects, there are three important risk scores to consider. The first is the *inherent* risk score, the second is the *residual* risk score, and the third is the *secondary* risk score.

[30] PMBOK Section 11.3.2.3 page 424

Inherent risk is the risk that exists before any controls or other mitigating factors are implemented (the gross risk or risk before risk response plans are implemented). This is the risk score calculated from the probability and impact before you've done anything to affect them.

Residual risk is the risk that remains after you've implemented a risk response plan. It's the risk remaining after you've taken action to remove the source of the risk, change the consequences, alter the probabilities, transfer the risk, or accept the risk.[31] For example, if your risk response plan for a risk was to mitigate the risk, your residual risk will be the risk that is left over after you have reduced the probability and/or reduced the impact.

Secondary risk is a risk that arises as a result of the risk response plan that you enact for the inherent risk. It is the new risk that has *been introduced* by enacting your risk response plan. For example, if your risk response plan for a risk is *transfer* by engaging a third-party vendor to do a portion of work, you are introducing a secondary risk that the third-party vendor may fail to meet their contractual obligations.

The inherent risk score is the risk score at the time of the workshop, before any control is applied or anything in the risk response plan is implemented to address the risk.

Note: If the risk response plan for the risk is *accept* then the inherent risk score will be the same as the residual risk score.

5.4 Risk Categorization

Risk categories help you to group the risks. Categories are useful if you want to filter your risks. The categories you use for your project's risk management should have been determined in the risk management plan.

I usually keep the categories high-level, concentrating on the project constraints: cost, scope, quality, schedule, resources, and benefits, but feel free to use different categories if you have categories that are more suitable for your project.

5.5 Cognitive Biases

The perception of risk feeds biases into the evaluation of identified risks. We need to identify bias and correct for them.

There are two factors that you should be cognizant of when assessing the risks that you have identified. The first is *groupthink* and the second is *optimism bias*. We discuss both of these biases below, and we outline how you can try to correct for them.

[31] http://praxiom.com/iso-31000-terms.htm

5.5.1 Groupthink

"If everyone is thinking alike, someone isn't thinking."

George Patton Jr., American General

Groupthink is the first bias you should try to counteract when soliciting the opinion of your participants. Wikipedia defines groupthink as

> *"Groupthink is a psychological phenomenon that occurs within a group of people in which the desire for harmony or conformity in the group results in an irrational or dysfunctional decision-making outcome. Group members try to minimize conflict and reach a consensus decision without critical evaluation of alternative viewpoints by actively suppressing dissenting viewpoints and by isolating themselves from outside influences."*[32]

Essentially, this means that people in groups tend to minimize tolerance of marginal thinking or viewpoints. Given that many of the dangerous risks are the ones that aren't obvious and are on the margins of common thinking, groupthink presents serious challenges to the facilitator of a risk workshop.

When working in groups, people tend to minimize conflict by suppressing opinions and ideas that differ from those of the group. Groupthink in risk workshops can narrow the band of risks that you identify and can warp the assessment of the probability and impacts of those risks.

Groupthink can cause people to suppress risks that may be contentious or risks that touch on political sensitivities, and groupthink can skew the assessment of risks' impact and probability.

We discuss how we can offset groupthink in section 5.6.3.

5.5.2 Optimism Bias

Optimism bias is the second factor to consider when compiling the participants' input into probability values for risks. Wikipedia defines optimism bias as "*a bias that causes a person to believe that they are less at risk of experiencing a negative event compared to others* "[33]

> Optimism bias – this is the proven tendency for appraisers to be too optimistic about key project parameters, including capital costs, operating costs, project duration, and benefits delivery. Over-optimistic estimates can lock in undeliverable targets, and it is therefore critical to make adjustments for this.
>
> The HM Treasury's *Green Book* recommends applying specific adjustments to correct for optimism bias at the project level. Cost estimates are increased by a set percentage to reflect evidence of underestimation from previous similar interventions. Adjustments should be based on an organization's own evidence base for historic levels of optimism bias.[34]

[32] https://en.wikipedia.org/wiki/Groupthink
[33] https://en.wikipedia.org/wiki/Optimism_bias
[34] Page 6, HM Treasury Green Book

If you ask your participants how likely each risk is to occur, optimism bias will cause many of your participants to reply with an unrealistically low probability for negative risks.

Optimism bias will skew your risk scores downwards for threats. If you take a simple average of all the participants' probability values, you will end up with an optimistic probability for your risks. When this optimistic probability value is multiplied by the impact, your risk score will also be unrealistically low.

There are many ways of mathematically weighing probabilities upwards to offset this optimism bias. Some of these methods are complex. There is no standard method of quantitatively determining the optimism bias for a particular group (at least, none that I've been able to find). So, if you don't have empirical evidence, I rely on a simple heuristic method to correct for optimism bias that I describe in section 5.6.4.

5.6 Qualitative Risk Analysis Process

When analyzing risks, I recommend working through the following steps:

1. Assess the impact that the risk would have on your project objectives if it were realized
2. Assess the probability of the risk occurring
3. Calculate the risk score
4. Record the inherent risk score in the risk register

5.6.1 Assess Impact

In risk management, the words *impact* and *consequence* are interchangeable. For the purposes of this book, I use the word *impact* to describe the effect a risk would have on the project if it were realized.

In assessing the impact of any particular individual project risk, we are trying to determine what effect it would have on the project objectives if the risk was realized (i.e., if the event we described as a risk actually occurred).

For example, if "the delivery of key components is delayed" is a risk event, then we need to try to determine what effect or impact a delay of the delivery of the key components would have on the project objectives.

Some risks are going to have impacts that are binary, i.e., they only have two states. Either they occur, and they have a set impact, or they don't occur, and there is no impact. For such risks, there is no scale of impact. An example might be the risk that a sponsor approves a document. There is no "partial" approval. Your sponsor either has approved the document or he has not approved the document; there is no halfway between the two states.

The impact of other risks may have a sliding scale. The impact could be small, or it could be large, or it could be any point in-between. In the case of such risks, it is safer to describe an impact on the high end of the impact scale. For example, consider the risk of one of your team members getting sick.

They could be sick with a sniffle and absent from work for a day. Or, they could be sick with a more serious ailment and be absent for a month. Or, they could be absent for any timeframe in-between.

If you have a risk that a key SME might be off sick, they might be sick with the common cold. If they only have a cold, they may be able to answer urgent questions and take care of some important tasks from home. And, if it is just a cold, they may only be absent from the workplace for a couple of days. That would lessen the impact of the risk.

But, if they come down with bronchitis, they may be unable to answer even the most urgent questions. And they may be absent for a month, unable to contribute to the project at all. That would increase the impact of the risk.

When it comes to assessing the severity of the impact of a risk that has such a sliding scale of impact, it is safest to assume an impact at the higher end of the scale. In the case of a risk of team members getting sick, it is safest to be pessimistic and assume they'll be absent for a longer duration. Assume they're going to come down with bronchitis, not the common cold.

Your organization may have a scale for categorizing each risk impact, and you may have guidance on what impacts should sit within each category of the scale. That scale is likely to tie directly to the risk matrix. If your organization does not have a scale, you would be well advised to create your own so that risk impacts are assessed consistently within your project.

The categories in your impact scale should be assigned a number in order of severity of impact — starting with the lowest impact as the lowest number.

Most organizations break impact down into three to six categories to describe the severity of impact. There is no hard constraint on the number of categories. If your organization doesn't have a standard and if you don't have a preferred practice, I recommend using five. You can see a simplified scale for categorizing impact in the example Impact Assessment Guide below.

Table 7 – Impact Assessment Guide

Impact Assessment Guide				
No.	Impact Severity	Cost	Schedule	Quality
5	Very High	>$1 million	Delay >6 months	Very Significant impact on overall functionality
4	Major	>$250,000	Delay >2 months	Significant impact on overall functionality
3	Moderate	>$100,000	Delay >1 month	Some impact on key functions
2	Minor	>$30,000	Delay >2 weeks	Minor impact on overall function
1	Negligible	<$30,000	Delay <2 weeks	Minor impact on secondary functions

In the simplified examples in the table above, we have only used the impact categories for cost, schedule, and quality, but you will likely want to include other categories as well. Some suggested categories are

- Schedule – the impact the risk would have on your schedule if the risk was realized; this usually forms delays (although not necessarily)
- Quality – the quality of some aspect of the project would be impacted; this might be tied to defects or some other quality measure
- Cost/financial – the financial impact of a risk being realized; most often, this relates to cost overruns, but it could include cash flow impacts, impacts on budget approvals, impacts on crossing the organization's financial year, capital expenditure vs. operational expenditure impacts, etc.
- Resources – impacts on assets, staff retention, staff morale, physical impacts on staff, external stakeholders, etc.
- Environment and community – impact on the environment or community; there may be a service your project introduces, replaces, or changes that could have a broader positive or negative impact on the community
- Management – risks that are realized can increase demands on management resources, which reduces their time available for their other responsibilities
- Reputation – potential impact on the reputation of the organization can have cascading impacts on sales, staff retention, staff recruitment, etc.
- Legal – impacts may include litigation, fines, or other penalties for breaches of legislation
- Benefits – how would these risks impact the benefits that the project business case is based on? Consider both financial and non-financial benefits
- Scope – how would these risks impact the project scope? Could they necessitate an expansion in the scope? Could they prevent the delivery of some elements of the scope?

Note: This is neither a prescriptive nor an exhaustive list. Your organization or industry may not consider some of these categories, or you may want to add extra categories. This list is provided only as an example.

As I introduce the topic of impacts in the workshop, I like to run through the impact categories with the participants to ensure everyone is clear on the categories. You will find that the impact categories help people think more consistently when assessing the impacts of risks.

When addressing impacts in a workshop, I like to keep a list of impact categories posted on the wall for the participants to consider. An impact table will give you an objective reference point to confirm the impact levels that people suggest.

For example, when someone suggests that a schedule risk has a moderate impact, you can say "So, the impact is between 1- and 2-months delay?" You'll be surprised how often people change their impact assessment when you refer them back to the reference table!

It is most effective to run through each risk and identify both the impact and the probability at the same time. I recommend a method of how to do so in the following sections.

5.6.2 Probability Analysis

The terms *likelihood* and *probability* are interchangeable as they relate to risk. ISO31000:2018 argues for the use of *likelihood* due to the mathematical association that the word *probability* has in the English language. But, both PMI and Axelos use the word *probability*.

For the purposes of this book, I'll use the word *probability* to refer to the chances of a risk event occurring.

In assessing the probability of a risk, we are trying to determine what the chance is of the risk being realized (i.e., the chance of the event we described as a risk actually occurring). If "the delivery of key components is delayed" is a risk event, then we need to determine what the chances are that there is a delay in the delivery of the key components.

Probability is most often broken down into four to six categories. There is no prescriptive number of categories, but I recommend using five, as per the diagram below. The categories should be assigned a sequential number in order of probability, starting with the least likely as the lowest number in the list. A five-category example of probability categories can be seen in the Probability Assessment Guide below.

Table 8 – Probability Assessment Guide

Probability Assessment Guide			
Score	Category	Probability	Impact
5	Almost Certain	>80% chance of occurring	The impact is expected to occur within the timeframe of the assessment
4	Likely	51-80% chance of occurring	The impact will probably occur within the timeframe of the assessment
3	Moderate	21-50% chance of occurring	The impact may occur within the timeframe of the assessment
2	Unlikely	1-20% chance of occurring	The impact is unlikely to occur within the timeframe of the assessment
1	Rare	<1% chance of occurring	The impact is highly unlikely to occur within the timeframe of the assessment

There are many methods of establishing the probability for each risk. If you have empirical data that you can use, then start with that. If your project will be reliant on a particular supplier, then their history of reliability may inform your probability.

Consider an example where your supplier has delivered late on eight of their last ten engagements. If they haven't addressed the problems that caused their tardiness, you could assume there is an 80% probability of them delivering late.

If you do not have empirical evidence to refer to, then you will be dependent on the expert judgement of your workshop participants.

5.6.3 Offsetting Groupthink

In Section 5.5.1 we explored the dangers groupthink poses to your assessment of impact and probability. To avoid the pitfalls of groupthink, I like to use anonymous voting.

Anonymous voting can be done in many ways, but one effective method is to give out a risk assessment sheet to each participant and ask them to record their estimate of probability and impact for each risk on the sheets. Then you can run through the risks one by one.

This method allows every participant to record their thoughts before they come under the influence of the thoughts of others. It is similar to the method we use to avoid the pitfall of first mover advantage through individual brainstorming, as described in in section 4.5.1.

An assessment sheet would typically provide a structured list with columns for

- Risk Number
- Risk Description
- Probability and
- Impact

I find it helpful to include a probability assessment guide and an impact assessment guide with the assessment sheet.

You can download free *probability and impact assessment sheets* at www.ictrisk.com.

After your participants have completed the exercise, collect the sheets so you can collate the probability and impact for each risk. If you're not going to create risk response plans in the same session, then collate the sheets after the workshop, as your analysis will not require any further input from your participants.

5.6.4 Offsetting Optimism Bias

In Section 5.5.2, we investigated the dangers that optimism bias poses to your assessment of probability. If your organization has an established method to deal with optimism bias, use that.

While the HM Treasury's *Green Book* provides some guidance in offsetting optimism bias for projects estimates, we don't have any industry metrics that can be used to offset optimism bias at an individual risk level.

If your organization does not have an established method to deal with optimism bias, you can use a simple heuristic method to offset optimism bias in estimating individual project risk probabilities.

This simple method to offset optimism bias starts by ranking your participants' probability values from highest to lowest. Then, take a simple average of the highest 75% of the group's probability scores (i.e., take a simple average of the 75% of the group that have the highest probability rating. Disregard from your calculation the most optimistic 25% participants' input).

This will give you a probability that is based on a large set of inputs but has been negatively weighted to offset the optimism bias. An example can be seen in Table 9, below.

Table 9 – Offsetting Optimism Bias

Original List		Ranked & Weighted List	
Name	Probability Score	Name	Probability Score
Sarah	3	Helen	4
Dale	2	Paul	4
Helen	4	Sarah	3
Diane	2	Nishant	3
Nishant	3	Lisa	3
Lisa	3	Lilia	3
Paul	4	~~Dale~~	~~2~~
Lilia	3	~~Diane~~	~~2~~
Unweighted Average	3	Weighted Average	3.33

Note: The 25% discount has been heuristically determined through my own past experience. As such, it is subjective. Feel free to determine a value that is consistent with your organization's or your personal experience.

As you can see in the table on the left above, the original unweighted average probability was 3. This value is low due to participants' natural optimism bias.

If we look at the table on the right above, we can see the ranked and weighted list. When we list the probability responses from high to low and remove the most optimistic 25% from the calculation (i.e., two out of our eight responses, Dale and Diane), we get a weighted average of 3.33. Our treatment to offset optimism bias has moved the probability score from 3.00 to 3.33.

This is a deliberately simplified example to show the concept. But it is easy to apply the same principle to more complicated situations to offset the optimism bias.

5.6.5 Rank

Now that you've established a value for the impact and one for the probability, you need a method of combining the two factors to establish a meaningful score for the risk. Provided you have ranked the impact and probability in the order I described earlier, you can simply multiply the impact value by the probability factor to obtain a risk score.

Impact x Probability = Risk Score

This is not a perfect system, but it does provide a quantifiable method of comparing a low-impact/high-probability risk with a high-impact/low-probability risk. Risk scores allow us to compare apples to oranges. It provides us with a calculated risk score that we can use to determine which risks we will prioritize. In doing so, it allows us to take calculated risks.

> *"Take calculated risks. That is quite different from being rash."*
>
> General George Patton, American General

Consider the example in Table 2, below. Risk A's probability is very likely (probability value of 5 – *almost certain*) but risk A has an insignificant impact (impact value of 1 - *negligible*). On the other hand, risk B is less likely to occur (probability value of 3 - *possible*) but risk B would have more of an impact if it did occur (impact value of 2 - *minor*).

When we use a risk score, we can compare and prioritize these two very different risks. With a risk score of 5 (1 x 5), risk A emerges as a lower priority than risk B, which has a risk score of 6 (2 x 3).

Quantifying the risk score provides us with a useful method to compare individual risks against each other and to prioritize them. You can see a summary of the comparison in the table below.

Table 10 – Risk Score Comparision

		Impact				
		1 Insignificant	2 Minor	3 Moderate	4 Major	5 Very High
Probability	5 Almost Certain	5 - Risk A				
	4 Likely					
	3 Possible		6 - Risk B			
	2 Unlikely					
	1 Rare					

5.6.5.1 Pareto Prioritization Analysis

You want to focus your attention on identifying the most important risks. But, how do you decide which risks you need to deal with first? And, are some risks caused by the same underlying issue?

The tools and techniques used up to this point will have answered those questions for most of your risks. But, for any risks that may have slipped through with those questions unanswered, we have Pareto analysis.

Pareto analysis is a simple technique for prioritizing possible causes by identifying the risks that will be resolved by treating these causes. By using this approach, you can prioritize the individual changes that will most improve your aggregate risk profile.

Pareto analysis uses the Pareto principle—also known as the "80/20 rule"—which was first coined by Italian economist, Vilfredo Pareto, in 1896 in his book, *Cours d'économie politique*. Pareto found that 20% of the causes generate 80% of the results. Pareto was surprised to find out how common this ratio was and how it was applicable to so many different situations.

When we use this tool in prioritizing risks, we're trying to look for the 20% of the individual risks that will generate 80% of the uncertainty that your project is exposed to.

The six steps in Pareto prioritization are

Step 1 – Identify the risks (addressed in Chapter 3)

Step 2 – Identify the root cause of each risk (addressed in Chapter 3)

Step 3 – Score the risks (addressed in Chapter 4)

Step 4 – Group the risks together by root cause (new activity)

Step 5 – Add up the scores for each risk group (new activity)

Step 6 – Prepare your risk responses (addressed in Chapter 6)

Obviously, we have already covered step 1 (identify the risks) and step 2 (identify the root cause of each risk). And, we have covered step 3 (score the risks) in this chapter. Step 4 (group the risks together by root cause) may be covered off by your Ishikawa diagram efforts in 4.5.3, but Step 4 and Step 5 (add up the scores for each risk group) are the new steps at this stage.

Step 6 (prepare risk response plans) will be addressed later in Chapter 6.

The Pareto prioritization should not be used to exclude the lower 80%. It is a simple tool that allows you to identify the most important 20% of the individual risks. These are the risks that contain 80% of the uncertainty. This technique highlights that you should apply more time and effort to the top 20% of risk scores.

5.6.6 Risk Visualization

5.6.6.1 Risk Matrix

A risk matrix is a tool that helps stakeholders visualize the impact and probability of a project's individual risks. It makes it easy for stakeholders to quickly assess a project's exposure to individual risks. A risk matrix allows easy and quick comparisons between risks with different characteristics.

In the example below, each cell within the risk matrix shows the risk score (i.e., the impact value multiplied by the probability value). I have included the risk scores for reference. The cells in the risk matrix that you would use for reporting are usually populated with the number of project risks that sit in each category.

Table 11 – Risk Matrix Showing Risk Scores for Each Cell - Threats

		Threat Impacts				
		Insignificant	Minor	Moderate	Major	Very High
Probability	Almost Certain	5	10	15	20	25
	Likely	4	8	12	16	20
	Possible	3	6	9	12	15
	Unlikely	2	4	6	8	10
	Rare	1	2	3	4	5

Table 12 – Risk Matrix Showing Risk Scores for Each Cell - Opportunities

		Opportunity Impacts				
		Very High	Major	Moderate	Minor	Insignificant
Probability	Almost Certain	25	20	15	10	5
	Likely	20	16	12	8	4
	Possible	15	12	9	6	3
	Unlikely	10	8	6	4	2
	Rare	5	4	3	2	1

Note: The risk matrixes above use the same colors to denote the risk score for both threats and for opportunities. At one end of the scale, the color red indicates something we want to action urgently, whether it be a dangerous threat or an enticing opportunity. On the other end of the scale, the color blue indicates something we are almost indifferent about, whether that be a low priority threat or a low priority opportunity.

The risk matrix will reflect the number of severity categories you use to categorize the risk scores. In the example above, the ranking score is reflected in five severity categories, which are usually color-coded to reflect the severity of the risks.

There is no hard and fast rule on the number of severity categories to use; some organizations use three simple RAG categories (red, amber, green).

In the example above, I have used the following five severity categories:

- Blue – Risk score of 1-4 = tolerable
- Green – Risk score of 5-8 = tolerable with periodic review
- Yellow – Risk score of 9-14 = tolerable with regular review
- Amber – Risk score of 15-19 = intolerable without treatment
- Red – Risk score of 20 and above = intolerable

How your organization categorizes the risk scores in the risk matrix will reflect your organization's risk appetite. Regardless of how your organization uses them, risk matrixes provide a much richer method for comparing and communicating risks than simply using the risk score on its own.

Table 13, below, is an example of a risk matrix that shows the number of individual project risks within each cell, as used for reporting the individual project risks.

Table 13 – Risk Matrix Showing the Number of Individual Project Threats that Populate Each Cell

		Threat Impacts				
		Insignificant	Minor	Moderate	Major	Very High
Probability	Almost Certain	3	4	1	0	0
	Likely	3	2	2	0	0
	Possible	2	0	1	2	1
	Unlikely	6	2	3	4	2
	Rare	4	1	4	1	1

Table 14 – Risk Matrix Showing the Number of Individual Project Opportunities that Populate Each Cell

		Opportunity Impacts				
		Very High	Major	Moderate	Minor	Insignificant
Probability	Almost Certain	0	1	1	2	0
	Likely	0	2	1	0	0
	Possible	1	1	0	2	2
	Unlikely	0	0	0	1	0
	Rare	1	0	3	1	1

The matrix example above is typical of matrixes used for reporting in projects. This matrix reflects the number of residual/secondary risks in each cell.

In this example, you'll note that there are no risks in the red cells for threats. This is because most organizations will not tolerate such high risk scores. Most organizations mandate that risk response plans are implemented for high score risks to reduce the residual/secondary risk scores to tolerable limits.

You'll also notice that there is only one risk in the red cells for opportunities. This isn't because of any actions the project team has taken. It merely reflects the rarity that projects find themselves in such fortunate situations!

Over the last decade, risk management has increasingly emphasized the identification of positive risks. But, you'll notice that the number of threats far outweigh the number of opportunities (49 to 20). This reflects the residual emphasis that most projects still apply to threats over opportunities.

Most projects I've worked on amalgamate the risks from each of the project objectives that they are using (e.g., cost, scope, quality, and schedule). But, if your project board or sponsor values each of the project objectives differently, you may want to augment that approach.

You may want to produce a separate matrix for each one of your project objectives. That would show how the different individual risk profiles compare for the different project objectives.

If you used the analytical hierarchy process described in section 2.2.1, you should have quantified priorities that you can put against each of the project objectives. If you have these quantified weightings, and if you have categorized your risks by objective, you could reflect that ratio of priority for each project objective.

You might choose to augment your risk visualization with tools that depict the different priority weighting across the different project objectives. You could use tools like the pie charts below in which the surface area of the pie has been adjusted to reflect the sponsor's appetite for risk in each of the different project objectives.

Diagram 25 – Reporting by Project Objective

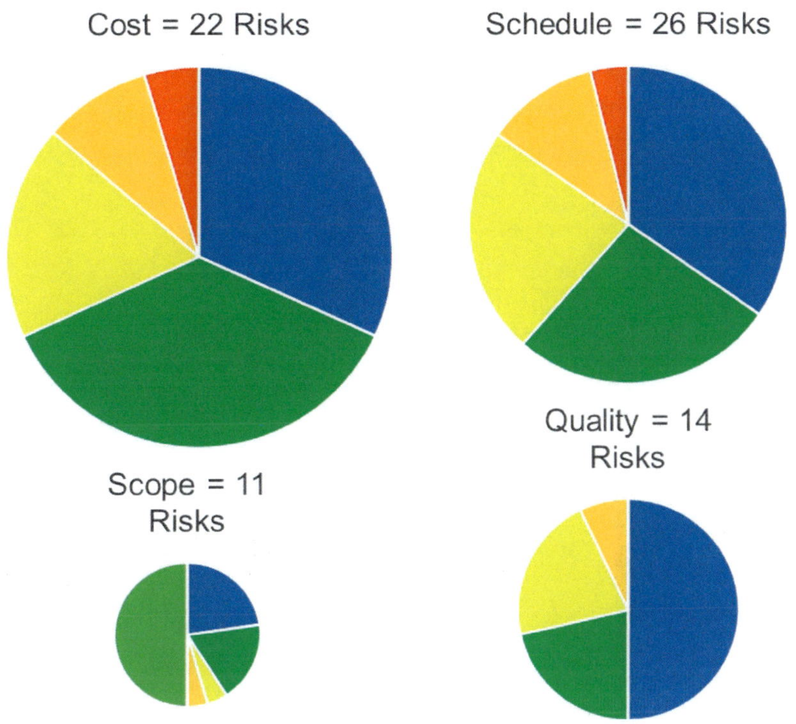

In the example above, we can see the number of risks for each project objective. We can also see how many of those risks are in each risk severity level (red, amber, yellow, green, and blue).

In the example above, the size of the pies tells us that cost is the most important project objective. The next most important objective is schedule, followed by quality. And then scope is the least important of the project objectives.

Each of your project objectives may have a very different risk profile as each project objective provides a different lens to view risk through. Depending on the nature of your project, this may be an important distinction to make for your project's decision making.

5.6.6.2 Bubble Chart

Another method of visualizing risk is to use a bubble chart. Bubble charts are more complicated than risk matrixes, and it takes more time to absorb the information they provide. But, bubble charts allow you to communicate more information than you can with a risk matrix.

A risk matrix is limited to the two axes that show probability and impact, the color indicating the categorization of the risk score and a number in each cell indicating how many risks have that particular probability/impact profile.

A bubble chart enables you to show more information than a risk matrix. A bubble chart conveys information through the vertical axis, the horizontal axis, the bubble size, the bubble quantity, the bubble shading, the bubble color, and the bubble pattern.

The use of bubble charts in risk management is less standardized than the use of risk matrixes.

You can slice and dice a bubble chart to visualize the information in the way that best serves your needs. You could even add extra variables if you want to convey more information (e.g., you could add different shapes). But remember that the more information that you add, the more effort and time it will take for people to understand the messages you are trying to communicate.

Diagram 26 – Bubble Diagram Example

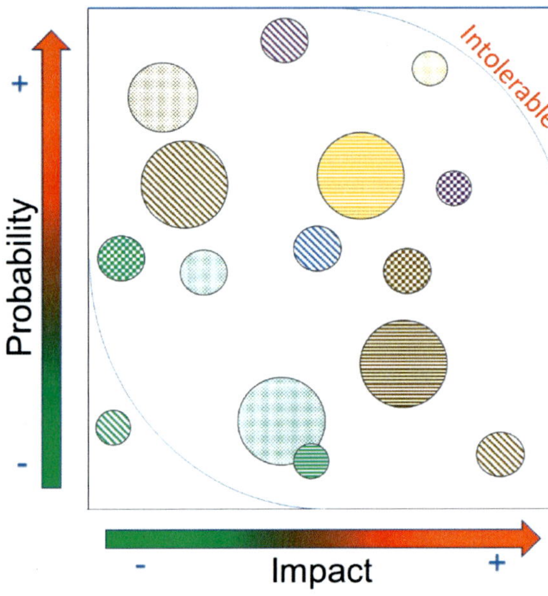

As you can see in the bubble chart above for negative risks (threats), the bubble chart can convey much more information than the risk matrix can. The bubble chart shows the individual project risk probability on the vertical axis, the impact on the horizontal axis, the proximity via the bubble size, the detectability via the shading, the risk response plan type via the color and the risk category by the pattern.

5.7 Chapter Summary – Qualitative Analysis

In this chapter, we've covered

1. Data quality analysis and how the quality of the data input into the risk evaluation processes determines the quality of the output
2. The extra parameters that you may choose to analyze if they are relevant to your project
3. The calculation of risk scores and the risk categorization of the risks that you have identified
4. What groupthink is and how it can affect the qualitative analysis of your individual project risks

5. What optimism bias is and how it can affect the qualitative analysis of your individual project risks
6. The qualitative risk analysis process
7. How to assess individual project risk impacts
8. How to assess individual project risk probabilities
9. How to offset groupthink
10. How to offset optimism bias
11. How to rank risks including the use of the Pareto prioritization analysis
12. Risk visualization and communication

6 Quantitative Analysis

Perform quantitative analysis is the fourth link in our risk management process chain of steps.

Diagram 27 – Quantitative Risk Analysis

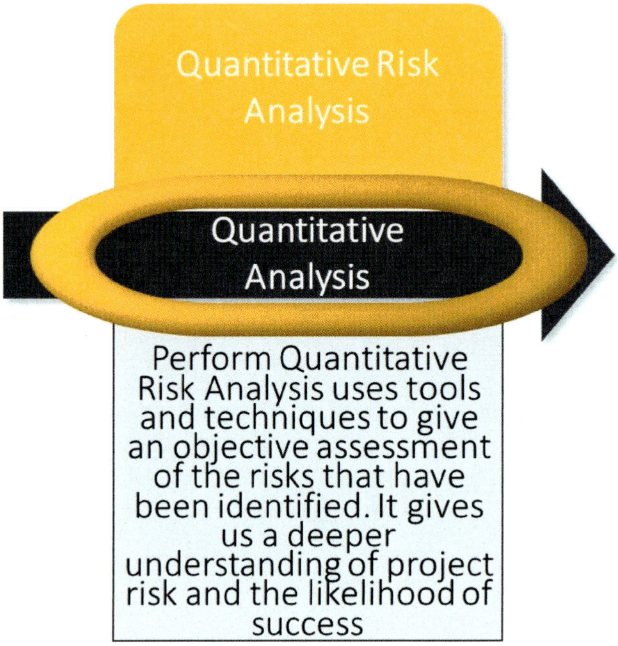

Quantitative analysis gives us our first chance to measure overall project risks. And, when we measure overall project risk, we are better able to manage that risk. And, when we're better able to manage overall project risk, we're able to make better informed project decisions.

> *"What gets measured, gets managed."*
>
> Peter Drucker, American management consultant, educator, and author

Measuring risks quantitatively enables us to manage risks more effectively. The risk profile of a project is more than just the sum of the risk exposure of its individual risks.

Qualitative risk analysis on its own can never give you a full view of a project's risk profile. The risk profile of a project can only be fully understood by performing BOTH qualitative and quantitative risk analysis.

Overall project risk can only be determined through solid, defensible, quantitative information about the likelihood and severity of each important risk faced, and that information can only be derived through quantitative risk analysis.

Without the information gleaned from quantitative risk analysis, one cannot verify whether the project is being run within the bounds of the sponsor's risk appetite. Nor is it possible to optimize the project's risk response plans and to compare the risk of different decision options with any meaningful level of precision.

If you are managing a project that is large, complex, or important, you should incorporate quantitative analysis into your risk management process.

6.1 What is Quantitative Analysis?

Performing quantitative risk analysis is the process of mathematically evaluating the collective effect of identified individual project risks on overall project objectives. The main advantage of this process is that it quantifies overall risk exposure. This process can also supply extra quantitative information that can be used to support risk response planning.[35]

I'll expand on that PMBOK concept to say that quantitative analysis gives us two main benefits. First, quantitative analysis enables us to measure overall project risk. Second, quantitative analysis enables us to calculate which individual risks have the most impact on overall project risk. Knowing which risks are most impactful enables us to decide to treat those risks according to their impact on OVERALL project risk.

Quantitative analysis gives us a chance to move from the subjective qualitative risk analysis to a more objective analysis. It gives us the chance to use quantitative metrics and to apply sophisticated software tools that perform calculations on those metrics.

Quantitative risk analysis simulations can be performed on cost, schedule, or a combination of both. In this book, I focus on a cost-based simulation, as it is the simplest way to convey the benefits that quantitative risk analysis simulations can bring to your project.

Obviously, we can only analyze the risks that we are aware of. Quantitative analysis is based on known risks. As such, the output from quantitative analysis necessarily excludes any unknown unknowns. The probabilities that we calculate in quantitative analysis cannot, and do not, include unknown unknowns. This caveat should be clearly made anytime we present the results of quantitative risk analysis.

Quantitative risk analysis simulations, such as Monte Carlo analysis, are both complex and complicated. In this book, we only aim to introduce the concept and outline the benefits of Monte Carlo simulations. It is beyond the scope of this book to provide enough information to enable the reader to perform the Monte Carlo simulations.

6.2 When to Use Quantitative Analysis

Both PMI and Axelos point out that a quantitative risk analysis is not mandatory for all projects.

---

35 PMBOK Section 11.4, page 428

Small, repetitive, or insignificant projects rarely merit the effort required to perform a quantitative risk analysis. If you are managing a series of small repetitive projects, quantitative risk analysis is likely not necessary.

For example, if you were running a project installing new printers across all your office locations, the project would not likely be important enough to justify the effort and cost of performing quantitative risk analysis.

But, most ICT projects are, by their very nature, complex projects. And, for the majority of ICT projects, the uncertainty that complexity introduces easily justifies the time and cost of performing quantitative risk analysis. If you're reading this book, you're likely involved in an ICT project that is far more complex than rolling out a series of new printers!

Quantitative analysis may not be mandatory for projects, but PMBOK recommends considering quantitative analysis if your project is any of the below

- Large in cost, scope, duration, or impact
- Complex – the nature of ICT projects is such that most ICT projects naturally score "high" on complexity
- Strategically important
- When there is a contractual requirement
- When there is potential for legal disagreements with vendors/suppliers
- When the board or sponsor needs assurance that best practices are being applied to the risk management of the projects that they are sponsoring[36]

If you are not sure if you should use quantitative analysis for your project, you can assess your project for free using our *Project Risk Profile Assessment* tool at: www.ictrisk.com.

The *Project Risk Profile Assessment* tool will help you assess your project to determine if quantitative assessment is appropriate for your project. The tool will help you understand the reasons why you should, or should not, consider using quantitative risk analysis.

Our *Project Risk Profile Assessment* tool is available for free for anyone who has purchased this book.

Our *Project Risk Profile Assessment* tool generates a printable report that you can take away for your consideration or to share. Your free project assessment can be found at www.ictrisk.com.

Quantitative risk analysis requires both specialized risk software and expertise to set up and run risk simulations. And quantitative risk analysis requires someone familiar with risk modelling to help with the interpretation of risk models.[37]

[36] PMBOK Section 11.4 Page 429
[37] PMBOK Section 11.4 Page 429

If you are fortunate enough to have someone within your organization with the expertise, tools, and capacity to perform quantitative analysis, then try to get them to help. If you don't have the tools or expertise in your organization, or if that expert doesn't have the capacity to help, consider hiring third-party project risk specialists.

Quantitative risk analysis is the costliest and most time-consuming aspect of risk management. As such, you should consider doing a cost-benefit analysis on including it in your project's risk management process. Due to the complexity of ICT projects, the cost-benefits almost always come out positive for projects of any significance. But, even if the outcome is almost predetermined, it's a good practice to do a cost-benefits analysis, and it provides evidence to support your decision in case that decision is challenged.

It is difficult to provide guidance on how much time and effort should be invested in risk management. The amount you invest will be dependent on your organization's and your sponsor's risk appetite, and on the ramifications of failing to achieve your objectives.

That being said, there is a growing trend in the industry to budget to spend 2% of the total project budget on second-line and third-line risk management.[38]

First-line risk management involves the day-to-day management and control activities. These include implementing project controls such as plans, schedules, and budgets to name a few. First-line risk management activities are often so tightly integrated with other project activities that they are difficult to segregate. As such, they are not included in the 2% estimate. First-line risk management costs are above and beyond the 2% estimate.

Second-line risk management involves defining, enforcing, and maintaining the standards and processes for project risk management. It includes activities that support the project by providing tools to manage new and emerging project risks, aggregating risks, and monitoring the effectiveness of project controls.

Third-line risk management involves independent assurance. It includes an independent assessment of the effectiveness of project risk management to provide the project board with confidence that the first and second lines of defense are operating effectively.[39]

If your project is strategically important, involves external partners, is complex, or is large in scope, duration, or impact, then you may decide to invest more than 2%.

If your project is not strategically important, doesn't involve external partners, is simple and straightforward, or is small in scope, duration, or impact, then you may decide to invest less than 2%.

When you do make decision on whether to include quantitative risk analysis and how much quantitative analysis to include, consider the research conducted by McKinsey and Oxford University as described in section 2.4. When we consider the underperformance of the typical project that their research uncovered, spending 2% of your budget on risk management seems a pittance!

---

[38] Oakes, G. 2008, 'Project Reviews, Assurance and Governance'

[39] Deloitte. 2017, Project Risk Management Applying the Three Lines of Defence Model to Project Risk Management

Many medium sized projects fall into the middle ground. Many projects decide to do quantitative risk analysis only once during their project. They rely on qualitative risk analysis for the rest of the project. This approach obviously limits the benefits of performing qualitative risk analysis, but it may be appropriate for budget-constrained projects that are medium in size and importance.

If you do only perform quantitative analysis only once, consider doing it before finalizing your business case. As you will see in the next section, quantitative risk analysis answers fundamental questions that will influence your sponsor's decisions on the business case.

6.3 Benefits of Quantitative Analysis

Quantitative risk analysis gives you valuable insights into your project that you cannot get from qualitative risk analysis alone.

Quantitative analysis enables you to calculate the probability of achieving cost and/or schedule objectives. Knowing in advance the likelihood of achieving your objectives enables more informed decision making and resource allocation.

But, the flip side is also true. Quantitative analysis also gives you the probability of losing money on the project. Quantitative analysis gives you an insight into the likelihood of realizing a negative return on investment (ROI) on the project. Knowing the likelihood of a negative ROI at the outset enables better mitigations and better go/no-go decisions.

Quantitative analysis gives you the ability to calculate contingency reserve requirements to achieve a set likelihood of successful project delivery. If your sponsor wants to be 80% sure of being able to deliver the claimed project benefits, quantitative analysis enables you to calculate the contingency reserves that you need to give your sponsor that level of confidence.

Quantitative analysis gives you the ability to rank and assess the factors that cause the most uncertainty. This *sensitivity analysis* enables you to create plans to reduce or eliminate one or more of the most impactful factors to reduce the overall project risk.

Quantitative risk analysis is the only dependable way to evaluate overall project risk. Quantitative risk analysis enables this assessment by aggregating the effect of all individual project risks on project outcomes.[40]

Quantitative evaluation of schemes after implementation is vital for producing realistic estimates of optimism bias to be used in the future. Monitoring and evaluating will also support improvements in costs, benefits, and timing for use in appraisals.[41]

6.4 Monte Carlo Simulations

A Monte Carlo simulation is a broad class of computational algorithms that rely on repeated random sampling to understand the impact of risk and uncertainty on overall project risk.

[40] PMBOK Section 11.4 pg. 429
[41] The Green Book, pg. 92

When you develop any model that plans for the future, you have to make assumptions. These might be assumptions about the cost of a purchase or how long it will take to complete a certain task.

Because you are making projections into the future, the best you can do is estimate the expected value. Most of the time, you can't know with certainty what the actual value will be. But, based on historical data, expert judgement, or experience, you can determine an estimate.

While this estimate is useful for feeding into a model, it contains some inherent uncertainty and risk, because it is an estimate of an unknown value.

In most ICT projects, it is often more appropriate to estimate a range of values. For both time and costs, we can usually determine a range of values, from most pessimistic, to most likely, to most optimistic.

When we provide a range of possible values instead of a single value, we create a more realistic view of what might happen in the future. When a model is based on a range of estimates, the output of the model will also be a range.[42]

Monte Carlos simulations can tell you, based on your series of ranges of estimates, how likely the resulting outcomes are.

Quantitative risk analysis applies a Monte Carlo model that simulates the combined effects of individual project risks and other sources of uncertainty to evaluate their potential impact on achieving project objectives.

Quantitative analysis produces three artifacts that are of particular interest for project risk management. These artifacts are

1. A histogram
2. A sensitivity analysis produced tornado diagram
3. A cumulative distribution curve diagram (S-curve diagram)

These three artefacts provide valuable information that enables better project decision making. We go into more detail on each in the sections below. But first, let us discuss in further detail the range estimates that feed into the Monte Carlo simulations.

6.4.1 Inputs into Monte Carlo Simulation Three-Point Stochastic Estimates

Let's say we were going to run a Monte Carlo simulation based on a simple triangular distribution for costs. The first thing we would need to do is determine a three-point estimate for each of the costs. The three-point estimate will consist of the most optimistic estimate, the most likely estimate, and the most pessimistic estimate.

When asked how long it will take to complete a task, people naturally want to give a range instead of a single value. We inherently know that there are variables that might impact how long any task would take to complete.

[42] https://www.riskamp.com/files/RiskAMP%20-%20Monte%20Carlo%20Simulation.pdf

Subject matter experts (SMEs) want to give a range of estimates for tasks. They hate to give a single value as an estimate. But, most of the tools and templates we use in project management demand that we input single value estimates, so we push our SMEs to come up with a single value estimate.

The SMEs will know that they can't tell us exactly how long a task will take. They know that a task will take between "most optimistic" X and "most pessimistic" Z, which is the variance. But since the tools and templates we use don't support a range of values, the SMEs are forced to give a "most likely" value of Y, which sits somewhere between "most optimistic" X and "most pessimistic" Z.

Monte Carlo simulations provide us with a tool that feeds on three-point estimates. Regardless of where we are in the project lifecycle, when we gather estimates, we should gather three-point estimates: the most optimistic estimate, the most likely estimate, and the most pessimistic estimate.

Using three-point estimates for our task's cost and schedule estimates enables us to make a quantitative and objective analysis. And the output of that quantitative analysis allows us to make more precise and accurate assessments.

6.4.2 Outputs of Monte Carlo Simulations

If we feed the most optimistic estimate, the most likely estimate, and the most pessimistic estimate values for each task into a Monte Carlo simulation, we can run the thousands of simulations required to calculate how the many task variances aggregate into a holistic view of probability. Let me explain that in simpler terms.

When we use quantitative analysis tools like a Monte Carlo simulation, we can provide decision makers with much richer information. We can provide them with information that allows them to make more informed decisions. Given the known risks and the known risk variance, Monte Carlos simulations enable us to provide sponsors with answers to the following questions:

1. What is the probability of achieving at least our stated return on investment (ROI) or our scheduled completion date?
2. What is the probability of losing money on the project or of missing a delivery deadline?
3. What contingency reserves would I need to be 80% sure (or another percentage) of achieving stated cost or schedule objectives?
4. What are the variables that have the largest effect on the probability of the project's outcome?

Most PMs have too many other urgent and important tasks to learn the Monte Carlo simulation tools and how to interpret the simulation results. If a PM only uses Monte Carlo simulations a few times a year, it doesn't make sense for them to buy expensive modelling software and to spend weeks learning how to use it. A PM's time is most often better spent elsewhere.

Specialist consulting companies provide projects with risk management services. These services provide a whole suite of project risk management including the running of the quantitative Monte Carlo simulations, the calculation of the probabilities, and the interpretation of the results. If you are running a project that is large, complex, or otherwise high risk, why not contact us for a free discovery consultation on how we could help you at www.ictrisk.com?

Specialist Monte Carlo computer software runs the quantitative risk analysis simulation models several thousand times. The output of the simulation shows the range of possible outcomes for the project. The output includes

- **Histograms** are the first type of output. They provide an interesting insight into the probability distribution.
- **Tornado diagrams** are the second type of output. They are produced by the simulation's sensitivity analysis to help identify the factors that have the highest impact on your overall project risk. Tornado diagrams depict the overall project risks that cause the most uncertainty in your project.
- **Cumulative distribution diagrams** are the third type of output. They are the most valued output of Monte Carlo simulations as they show the interface between probability and delivery cost/schedule.

6.4.2.1 Histogram

Histograms are the first output from the simulation. Histograms present the number of iterations where a particular outcome resulted from a simulation.

In the example below, the left vertical axis shows the histogram values (i.e., the values for the bars in the bar chart). The vertical bars show the percentage for each point on the horizontal axis. They show the probability of the project delivering in any particular week.

You will notice in the example below that the histogram represents a standard distribution (bell curve) that has not been normalized.

Diagram 28 – Histogram

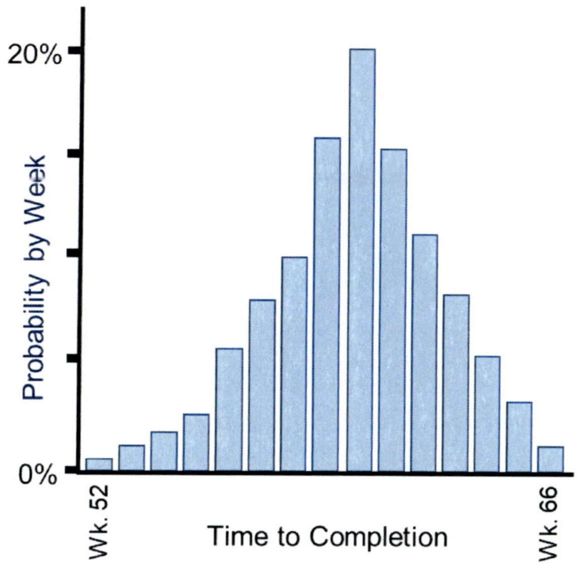

6.4.2.2 Sensitivity Analysis – Tornado Diagram

Sensitivity analysis is the second output from Monte Carlo simulations. Sensitivity analysis determines which overall project risks add the most uncertainty to your project.

Monte Carlo simulations enable sensitivity analysis. Sensitivity analysis tells us which factors have the greatest impact on the project risk. Sensitivity analysis produces tornado diagrams as an output.

The tornado diagram portrays the calculated correlation coefficient for each risk that can influence the project. Risks are ordered by descending strength of correlation. This gives the diagram the appearance of a tornado, hence the name "tornado diagram."

Tornado diagrams graphically depict the risk factors that cause the most impact on your project's overall risk profile. For an example, see the tornado diagram below.

In this example we can see that Risk 1 has the greatest impact on the project's risk profile. The next greatest impact on your project's risk profile is Risk 2, followed by Risk 3, etc.

Diagram 29 – Tornado Diagram

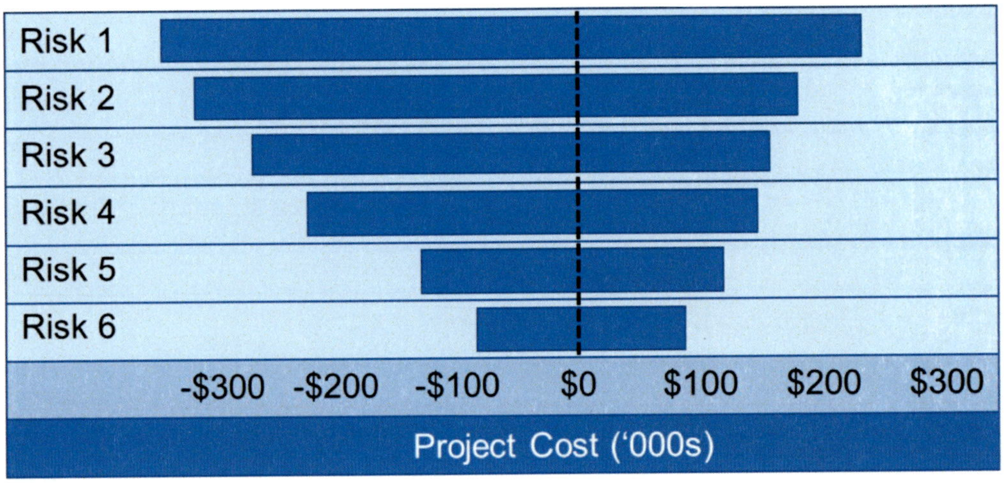

In this example, we can see that risks 1, 2, and 3, in descending order, are the most impactful risk factors for our project. If we find that the overall project risk profile is outside of our risk tolerances, we could apply a risk response plan to one or several of those risk factors. That would likely bring the overall project risk profile back into a tolerable range.

Once we have applied our risk response plan to one or more of the risk factors in the tornado diagram, we could re-run our Monte Carlo risk simulation. The Monte Carlo simulation will calculate our revised overall project risk profile to determine how much we have reduced it. Once we know how much we have reduced our overall project risk profile, we can assess if it is now within our project risk tolerance or if we need to implement more corrective actions.

6.4.2.3 The Cumulative Distribution Curve diagram (the S-curve Diagram)

The third output of our Monte Carlo simulation is the cumulative distribution curve diagram.

The cumulative probability distribution diagram shows the user the amount of time or the amount of cost most likely for any particular probability. This cumulative probability distribution diagram allows us to determine three critical pieces of project information.

- The probability of achieving the claimed ROI
- The probability of losing money on the project
- The contingency reserves that would be required to achieve a certain probability

These three critical pieces of information enable organizations to make better informed project decisions.

Diagram 30 – Cumulative Distribution Curve

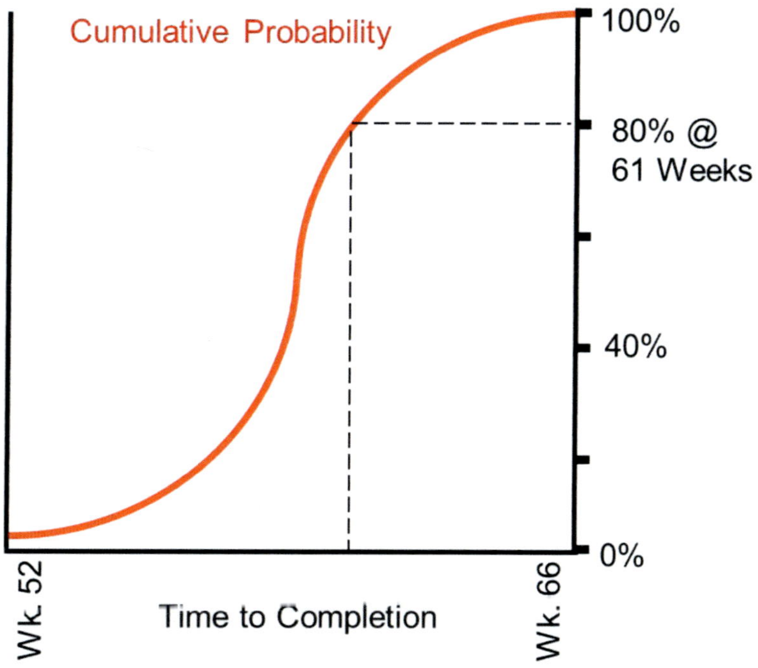

6.4.2.4 Combined Histogram and Cumulative Distribution Curve

In the example below, we have a cumulative distribution curve overlaying a histogram. In this example, the left vertical axis shows the histogram values (i.e., the values for the bars in the bar chart). The right vertical axis shows the cumulative percentages, the S-curve. The vertical bars show the percentage for each point on the horizontal axis. The S-curve shows the cumulative percentage value of each of the histogram bars as they are sequentially aggregated, one after the other.

You will notice in the example below that the histogram represents a standard distribution (bell curve) that has not been normalized.

Diagram 31 – Cumulative Distribution Curve with Histogram Overlay

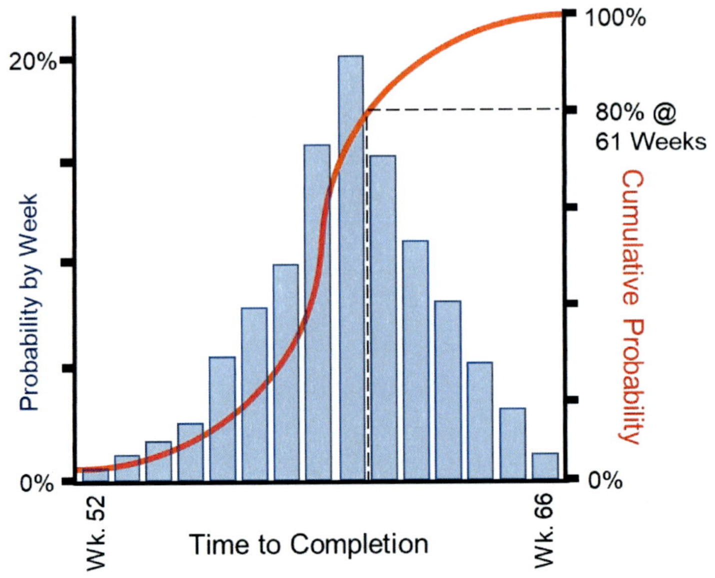

6.4.2.5 The Probability of Achieving the Claimed ROI

When we gather the known variance by getting our subject matter experts to provide three-point estimates for the costs of each task, we create an opportunity to run a Monte Carlo simulation. When we run a Monte Carlo simulation, we run modelling simulations that can determine the probability of achieving at least our claimed ROI.

When we provide the decision makers with the known probability of achieving the claimed ROI, they can make better informed decisions, and we can rest easy in the knowledge that the decision makers are supporting the project knowing full well the probability of achieving the claimed ROI. If the project delivers less than the target ROI, we can point out that the project was supported knowing the probability of achieving the target ROI.

6.4.2.6 The Probability of Losing Money on a Project

Depending on the organization and the industry, it is often just as important for decision makers to know the probability of losing money on a project as it is for them to know the probability of achieving the claimed ROI.

Losing money on an important project can cause serious, and even existential, problems for organizations. Careers can be damaged, and employment can be terminated when projects return a negative ROI. Failed projects can take down project sponsors, project managers, and whole project teams.

But, when we provide the decision makers with the known probability of returning a negative ROI, they can assess the risks and make better informed decisions.

And, when we perform a sensitivity analysis to tell them which factors have the greatest influence on the risk of achieving a negative ROI, then they can make decisions on how to treat those risks. Once those risks have been treated, we can run the simulations again to show how the known probabilities of returning a negative ROI have improved.

When we use Monte Carlo simulations to show the probability of achieving a negative ROI, we can rest easy knowing that decisions are being made with the best information available. We can proceed knowing that the decision makers are supporting the project knowing full well the probability of delivering a negative ROI.

If the project delivers a negative ROI, we can point out that the project was supported knowing full well the probability delivering a negative ROI.

6.4.2.7 Calculating Contingency Reserve Requirements

We may calculate a probability of achieving an ROI that the project sponsor finds unacceptable. If that is the case, we can also calculate the amount of contingency reserves that we require to bring the probability into a range that is acceptable to our sponsors.

For example, say we calculate the probability of achieving our stated ROI goal as 50%. If the project sponsor determines that 50% is too low and that they want 80% probability of success, we can calculate the amount of contingency reserves that would be required to achieve 80% probability of success.

When we provide decision makers with the probability of achieving their project goals, we give them the opportunity to tell us what minimum probability they would find acceptable. We can then use our modelling simulation tools to calculate the amount of contingency reserves that we require to achieve their desired probability of success.

6.4.2.8 Rank and Assess the Factors That Most Impact on Overall Project Risk

When we use our modelling tools to run a sensitivity analysis, we can show which risk factors cause the most variance. If we run the sensitivity analysis and show the sponsor how much of an impact each of the factors has on our probability, we can recommend how we can mitigate or hedge against one, or several, of those risks.

If we know the degree to which those factors impact the overall project risk, then we can mitigate or hedge against those factors. We can take action that can bring the project probability of success back into a range that the project sponsor is comfortable accepting.

For example, if we run a sensitivity analysis and we find that the variability of the interest rates over the next five years has the highest impact on the project uncertainty, we can recommend ways to treat the interest rate risk.

If our exposure to interest rate risk was causing unacceptably high overall project risk, we could enter into forward contracts or futures contracts or some other financial hedging service. We could do that for some, or all, of the projected impacted costs variances. Then we could add the costs of hedging and the reduced uncertainty to our model and rerun the model simulations. We could then show the sponsor the effect that the mitigation or hedge has had on the project's overall risk profile.

Thus armed, project sponsors and project boards can make decisions based on the best-known quantified information that we have at the time.

And the PM, project team and SMEs can rest assured knowing that they have provided the decision makers with a full disclosure and transparency of the variability caused by the known risks.

6.4.3 Quantitative Risk Analysis Metaphor

Running a project is kind of like driving a car in a rainstorm. Your view of where you're going and what's in your path is never going to be perfect when it's raining. But your view is much better once you turn the windshield wipers on!

Before you turn on your windshield wipers in a rainstorm, you have a poor view of the road ahead of you. The shapes you see are vague and blurry. It's difficult to judge which shapes pose a danger until they're very close.

Approving and running a project without quantitative risk analysis is kind of like driving in a rainstorm without the windshield wipers on. Not a good idea!

Using quantitative analysis is kind of like turning the windshield wipers on when driving in the rainstorm. Suddenly, you have a much better view of where you're going. You can see the dangers around you much more clearly. Instead of vague blurry shapes, you can see clearly what is coming your way, and you can see risks much earlier. You can more easily judge what dangers you need to avoid.

Driving in a rainstorm with your windshield wipers on is not perfect. Your vision is still impaired by the rain, but your vision is much better than if you didn't have the windshield wipers on. If you're going to drive in a rainstorm, you're going to want to use your windshield wipers any time you can!

So too with projects. If you want to have a better view of where you're going and if you want to see the dangers around you much more clearly, perform quantitative risk analysis. It will greatly improve your view of where you're going! It will let you spot risks earlier and see them clearer. Running a project with quantitative analysis doesn't give you a perfect view of your risks, but it gives you a much better view than when running a project without it!

> "*It is better to be roughly right than precisely wrong.*"
>
> John Maynard Keynes, British economist

6.5 Quantitative Methods to Apply to Risk Response Plans

There are quantitative tools and techniques that can be used to assist you in project risk management. These are alternative analysis tools and techniques that provide quantitative methods to assess options. They provide a quantitative metric to inform decision making. These tools are most often employed when assessing risk response plan options for individual project risks.

6.5.1 Influence Diagrams

The first alternative quantitative analysis tool is *influence diagrams*. We describe influence diagrams first, because they can be used in conjunction with the same Monte Carlo analysis that we discussed in the previous section.

Influence diagrams are related to decision trees and influence diagrams are often used in conjunction with them. Influence diagrams display a summary of the information contained within a decision tree. Influence diagrams are effective in communicating the dependencies amongst the different variables.

Influence diagrams do not contain the level of detail that a decision tree contains. But, influence diagrams are very helpful in explaining decisions that have been reached using a complex decision tree.

Diagram 32 – Influence Diagram Symbols

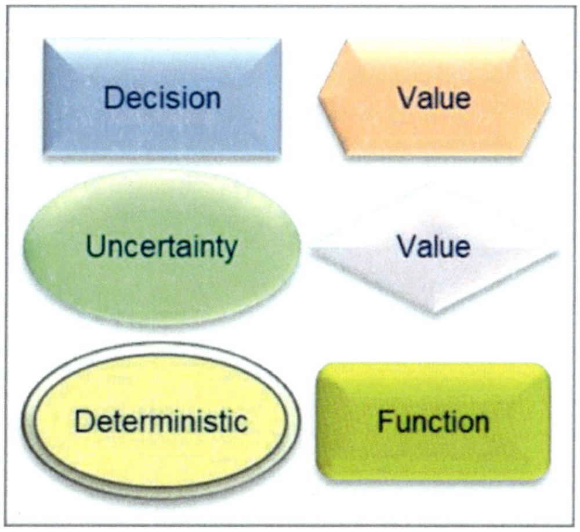

Diagram 33 – Influence Diagram Example of a Product Decision

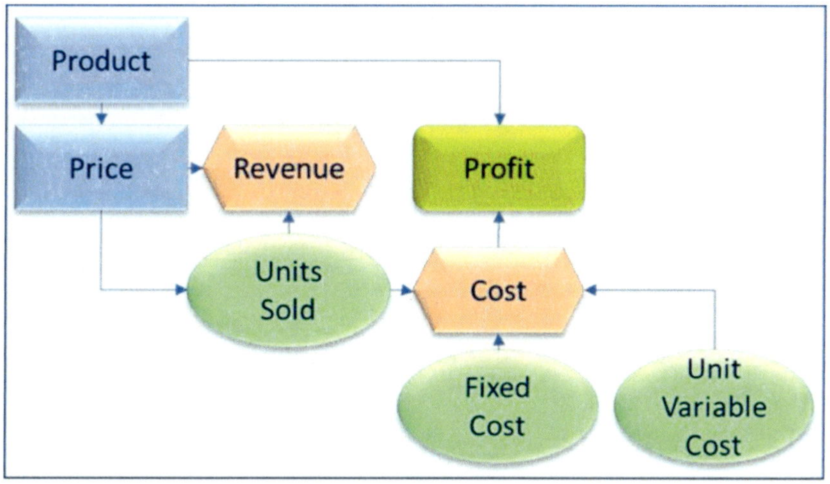

You can see in the above influence diagram example a situation where a decision needed to be made on product selection and the price assigned to the product. In this example, the number of units sold, the fixed cost, and the unit variable costs are all uncertain. The revenue and costs are both values that will be determined, at least in part, by the uncertainties, and profit is the function that leads to an outcome.

The software company SmartDraw provides more succinct instructions on how to draw influence diagrams than I could. According to SmartDraw,

> "There are many different ways to create an influence diagram.
>
> Here are the basic steps:
>
> - **Start your tree.** Draw a rectangle near the top-left corner. This will be the first 'node' or shape. In this rectangle, write the first question or a criterion that leads to a decision.
> - **Add to the process.** Referencing the decision tree, add more shapes to the page as needed, labeling each. A series of decisions, functions, and chances should lead to an objective.
> - **Add connectors.** Use a combination of straight or angled lines between all related boxes with arrowheads indicating the flow of influence. Each box should be connected to at least one line."[43]

When we add values and probabilities to our influence diagrams, we get a pictorial representation of the quantitative analysis we want to perform. We usually supplement this pictorial representation with a table below that holds the actual calculations used.

6.5.2 Multi-Criteria Decision Analysis (MCDA)

Comparing several different complex options without a set of criteria and tools to evaluate them would be very difficult. Without tools, our brains can only consider a very limited number of criteria. When you have several criteria and several options, the task of comparison becomes too overwhelming to work out in our heads.

Without a tool, when confronted with multiple competing criteria, we often end up relying on intuition to make a decision. We decide on the basis of a "gut" feeling. That is unlikely to produce optimal results. And, if we followed our intuition, it would be embarrassing if someone asked us to explain how we had come to our decision!

We need to take the time before making multi-criteria decisions to consciously and rationally determine the criteria we want to use to evaluate the decision. This will help us organize and track our assessment, and it can greatly reduce the effort required for the task. Establishing evaluation criteria for a MCDA also provides us with evidence of the rationale used in the assessment if someone asked us to explain how we made the decision.

The MCDA tool provides a reference point if any of our assessments are questioned. If someone were to ask why we chose the option we chose, we can show them the specific methods and calculations that we used to come up with our decision.

There may be an untold number of criteria that we could use to evaluate options. We need to determine which criteria are most relevant to the decision at hand, and we need to restrict our

[43] https://www.smartdraw.com/influence-diagram/

evaluation to those criteria. If we don't determine and restrict ourselves to the most important criteria, the amount of time required for the task could spiral out of control.

MCDA enables us to determine the set of gauges that we will use to measure our options. MCDA gives us consistent measurement points that allow us to assess our options according to our pre-determined criteria. MCDA provides us with a structured tool to compare apples to oranges.

We can use MCDA with either weighted or unweighted criteria. MCDA can be used with weighted criteria to accommodate for situations where one or more criteria is quantifiably more important than others. Or, MCDA can be used without weighting, for situations where all criteria are of equal importance.

Evenly weighted criteria would make our evaluation simpler. But, in most situations, there will be a distinct weighting where one or more criteria are clearly and quantifiably more important than others are. MCDA can be adjusted to accommodate for such weighting as required.

In the table below, we have a simplified example of MCDA for the assessment of four options against five criteria (cost, availability, experience, ability, and knowledge).

Table 15 – Multi Criteria Decisions Analysis

Option	Cost	Availability	Experience	Ability	Knowledge	Total Score Unweighted	Total Score Weighted
Weight	(x1)	(x1)	(x3)	(x1)	(x1)	n/a	n/a
Option 1	5	6	4	5	2	22	30
Option 2	2	8	3	8	9	30	36
Option 3	4	7	9	6	6	32	50
Option 4	1	7	7	9	8	32	46

You will notice in the example above that we've assessed criteria on a scale so that we can compare dissimilar metrics. If we didn't use a scale, it would be very difficult to compare a cost metric like $9,000 to an availability metric such as 30 days. Applying a scale enables us to compare apples to oranges.

For each variable, I've used a scale from 1 to 10. For cost, a value of 1 is the least favorable (i.e., most expensive) and 10 is the most favorable (i.e., cheapest). To do so, we have had to create "bands" to allow for meaningful comparisons (e.g., $9,500 to $10,000 is scored as "1" $9,000 to $9,500 is scored as "2," etc.).

Creating bands and scales is a subjective assessment, but it allows for a meaningful comparison across varied criteria. When creating your bands for your criteria, let the range of values for your options dictate the parameters for your bands.

The options in the above example have been evaluated to assess which option is most appropriate given the predefined criteria.

If we look at the unweighted score where each criterion was considered equally, it is a tie between option 3 and option 4, both with an unweighted score of 32. But, if we look at the *weighted* score, then option 3 emerges as the clear winner with a score of 50.

When we include the fact that experience is the most important criteria and we weight the ratings accordingly (3x for experience), then the results change significantly.

MCDA is a very useful tool for comparing multiple disparate and complex options. Wherever and however you decide to use it, MCDA is a versatile and valuable tool that will allow you to rationalize complex decisions.

6.5.3 Decision Trees

Decision trees are visual tools that you can use to support the selection of the best of several options. The tool enables you to show alternative paths as branches that represent alternate cascading decisions. The outcome from each of the decision paths can be found at the end of the decision branches.

A decision tree is a flowchart-like diagram that allows you to compare the various outcomes from a series of decisions. It is a helpful tool for decision-making and for planning. The primary advantage of using a decision tree is that it is easy to follow and understand.

Decision trees are usually oriented top down or left to right. Consider our simplified left to right example below.

Diagram 34 – Decision Tree Diagram

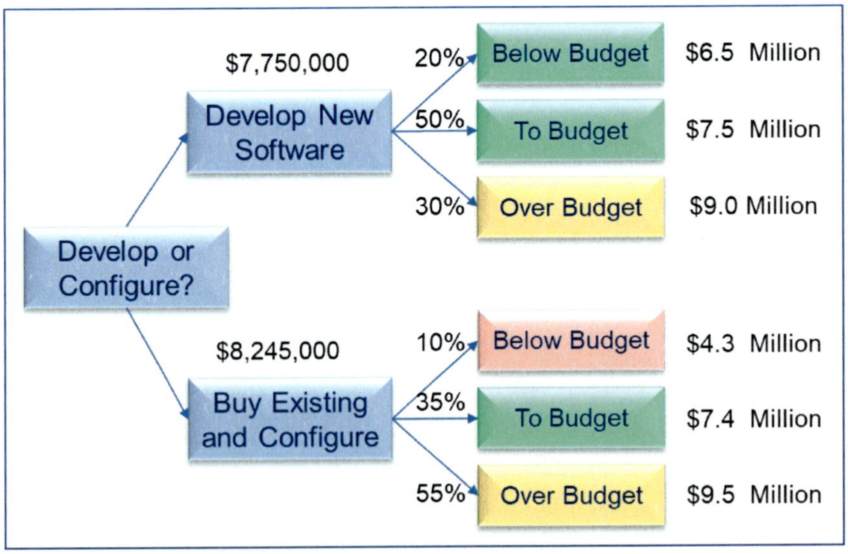

In our example above, we start with a decision whether to develop a software tool from scratch or to purchase an existing software tool and configure it to our needs. In this case, either option could result in one of three different outcomes.

Each of the three outcomes for the two different options has both a likelihood of occurring (expressed as a percentage), and an expected cost associated to it (on the far right of the diagram above) in dollar terms.

For each of the outcomes, we can multiply the likelihood by the predicted value for each outcome to come up with that outcome's contribution to the option's value.

So, for the "develop new software" option, we would multiply the "below budget" outcome likelihood (20%) by the "below budget expected outcome value" ($6.5M) to get an "outcome contribution" of $1.3M for the "below budget expected value." We can then do the same for the "to budget" option and the "over budget" options.

If we then add each of the three outcome's value for each option, we can come up with an estimated monetary value for that option. It may be clearer by running through our example in a table format below.

Table 16 – Decision Tree Calculation Table

Option	Outcome	Likelihood of Outcome	Value if Realised	Outcome Contribution	Options' EMV
Option 1 - Build	Below Budget	20%	$6,500,000	$1,300,000	
	To Budget	50%	$7,500,000	$3,750,000	
	Over Budget	30%	$9,000,000	$2,700,000	
	Build Total Expected Monetary Value				$7,750,000
Option 2 - Configure	Below Budget	10%	$4,300,000	$430,000	
	To Budget	35%	$7,400,000	$2,590,000	
	Over Budget	55%	$9,500,000	$5,225,000	
	Configuration TotalExpected Monetary Value				$8,245,000

In the above example, we can see that each option has three potential outcomes: "below budget", "to budget," and "over budget." For each of those outcomes, there is a likelihood for that outcome being realized. If we multiply each outcome likelihood by each expected outcome value, then we get an "outcome contribution" value. If we then aggregate each of the three outcome contributions for each option, then we get the expected monetary value for each option.

As we can see in the above table, the expect monetary value for the build option is $7,750,000. As the expected monetary value for the build option is lower than for the configure option of $8,245,000, the decision tree indicates that the build option is the preferred choice.

There are many situations in projects where decision trees can be useful in project management. I have found them especially useful when assessing options for risk response plans. Hopefully, this simple example shows how the concept can be used to choose between alternative options.

6.6 Chapter Summary – Quantitative Analysis

In this chapter we have covered quantitative risk analysis. We have considered

- What quantitative analysis is
- When to use quantitative analysis
- The benefits that quantitative analysis can bring to your project
- Monte Carlo simulations and three-point estimation techniques
- The outputs of Monte Carlo simulations, including histograms, sensitivity analysis / tornado diagrams and cumulative distribution curves

And, we've covered how Monte Carlo simulations can answer four fundamental project questions:

1. What is the probability of achieving the target ROI?
2. What is the probability of losing money on the project?
3. How much contingency reserve would be needed to achieve a predetermined level of certainty of success?
4. Which factors have the largest impact on overall project risk?

We've also run through quantitative tools and techniques that can be used to evaluate options for risk responses. These include

- Influence diagrams
- Multi-criteria decision analysis (MCDA)
- Decision trees

We can't, of course, calculate the probability of *unknown* risks. But we can, and should, calculate the probability of the *known risks*. Despite the time, effort, and cost of quantitative analysis, the knowledge we gain is worth it for all but the simplest or least-important ICT project.

> *"If you think knowledge is expensive, try ignorance."*
>
> Derek Bok, American lawyer and educator

If we have the ability to generate this knowledge to answer these fundamental questions, we do our projects a disservice if we continue with our projects without quantitative analysis. To not use quantitative risk analysis tools to answer the four fundamental questions is to do a disservice to the project sponsor, the project team, and to ourselves.

While qualitative risk analysis enables us to assess individual project risks, quantitative risk analysis enables us to assess four vital risk factors.

First, it enables us to assess the overall project risk to determine the likelihood of delivering the target ROI.

Second, it enables us to assess the overall project risk to determine the likelihood of achieving a negative ROI.

Third, it enables us to assess the overall project risk to determine contingency requirements we need to set aside to achieve a set probability of success.

Fourth, it enables us to numerically determine which individual risks have the biggest impact on our overall project risk, how big that impact is, and how we can reduce it.

7 Plan Risk Responses Process

Plan risk responses (also called “plan risk treatments”) is the fifth link in our risk management process chain of steps.

Diagram 35 – Plan Risk Responses

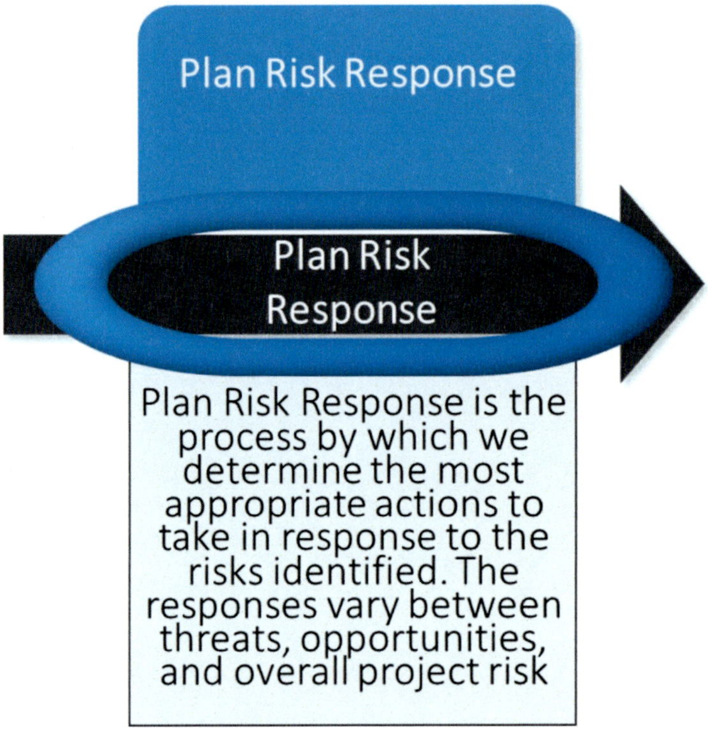

Now that we have thoroughly analyzed both overall project risks and individual project risks, we need to decide what we are going to do about them.

A risk response plan is necessary to address both individual risks and overall project risks. A risk response plan is the method we use to plan to manage each individual risk, and a risk management plan can describe how we are going to handle overall project risks.

Risk response plans can reduce individual threats, increase opportunities, and minimize overall project risk. Risk response plans help us prepare for the uncertainty that our projects will inevitably encounter.

> *"Expect the best, plan for the worst, and prepare to be surprised."*
>
> Denis Waitley, American motivational speaker, writer and consultant

The process of developing options and preparing plans on how to deal with the project risks is called the plan risk response process.

The plan risk response process not only determines what your project will do in response to risks, it also determines how and when you will execute the risk response plans. And, it determines who will be responsible for each risk response.

Individual risks each have a risk response plan, and each risk response plan has a risk owner who is responsible for the risk response plan. The risk owner is responsible for ensuring that the risk response is enacted for that individual risk.

You should also plan and implement risk response plans for overall project risks. The project manager is responsible for any risk response plan intended to address overall project risk.

7.1 Integrated and Iterative Risk Response Plans

The plan risk response process is both integrated and iterative.

The actions identified need to integrate with other project activities and documentation. At a minimum, risk response actions must integrate with the WBS, the schedule, and the budget.

Risk response plans should be integrated into the project plans, documents, and processes of the project, in consultation with appropriate stakeholders.[44]

The risk response process also needs to be iterative so that the risk response plans are kept current to the changing project environment. Any updates to the risk response plans need to cascade into any related project documents that are impacted by the changes.

Risk response plan options are not necessarily mutually exclusive.[45] The relationship between risk and risk response plan is not necessarily 1-to-1. This means that you may have a risk response plan that covers several risks, and you may also have several risk response plans for one risk.

One risk response plan may have a broad and deep impact across the project. Such a risk response plan could conceivably cover off many risks. For example, you may have a risk response plan of transfer where you outsource much of the project work. Depending on the nature of the work outsourced, that risk response plan could conceivably cover off many risks across cost, quality, schedule, or scope.

Conversely, you may have a risk that has a broad and deep impact on the project and is important enough to merit several risk response plans. An obvious example would be a risk that merits both a risk response plan that reduces the probability of the risk occurring and a separate risk response plan to reduce the impact that the risk would have if it did occur.

[44] ISO 31000:2018 Section 6.5.3
[45] ISO 31000:2018 Section 6.5.2

7.2 Appropriate Risk Response Plans

Risk response plans should be appropriate for the importance of the risk that it is meant to treat. Important risks will likely merit well-developed, detailed risk response plans. Less important risks likely merit more succinct plans that do not require as much effort to create or implement.

As part of that appropriateness, risk response plans should be cost effective. You will have to consider your sponsor's risk appetite, but in most cases, it doesn't make sense to spend more on a risk response plan than the Expected Monetary Value (EMV) of the risk (EMV = expected cost of the risk multiplied by the probability of occurring).

For important risks, the most appropriate risk response plan may be a combination of response plans. You may have a primary plan to reduce impact. You might have a primary plan to reduce the probability. And, you might have contingency or fallback plans in case either of the primary plans fail to achieve their objectives. The composition of risk response plans depends on the nature of the risk and the importance of the risk.

Deciding on the most appropriate risk response plan, or mix of plans, may be complicated for larger projects.

If you find that deciding on the risk response plan is difficult, or if a risk is important enough that you want to apply some analytical rigor to its assessment, consider using some of the quantitative tools and techniques described earlier in this book. AHP, MCDA, decision trees, and influence diagrams can all help in deciding on the most appropriate risk response plan.

7.3 Impact on Residual and Secondary Risk Scores

The type of risk response plan you select will determine the impact that the risk response plan will have on any residual and/or secondary risk score.

For any of your risks with a risk response plan of "accept," the residual risk score will be the same as the inherent risks score.

For most of your negative individual risks, the residual and/or secondary risk score will be lower than the inherent risks score.

It is this reduced level of risk that should feed into contingency reserve planning. The funding that you set aside for contingency reserves to cover the known risks should be calculated to cover the residual and/or secondary risk, not the inherent risk. Contingency reserves should be calculated using EMV techniques as described in 3.4.2.1.

7.4 Risk Evaluation

Part of the risk response planning process is risk evaluation. The purpose of risk evaluation is to support risk response plan decisions.

Risk evaluation involves comparing the results of the risk analysis with the established risk criteria to determine where additional action is required.[46]

Risk evaluation isn't just quantitative and qualitative analysis. It also extends into risk response planning. It includes evaluating the risks to identify what the most appropriate risk response plan would be.

The approach to risk evaluation in risk response plans may vary from one organization to the next. Attitudes to risk depend on the industry you are in, the maturity of the organization, and the cultures in which you are operating.

For example, an insurance company that has been operating for a hundred years is likely to have a much more conservative attitude to risk than an Internet start-up that has existed for six months.

7.5 Types of Risk Response Plans for Threats, Opportunities, and Overall Project Risk

The risk response plans for threats, opportunities, and overall project risk are distinct but related. These differences and relations can be seen in the table below.

Table 17 –Types of Risk Response Plans Table

Threats	Opportunities	Overall Project Risk
Escalate	Escalate	Avoid
Avoid	Exploit	Exploit
Transfer	Share	Transfer/Share
Mitigate	Enhance	Mitigate/Enhance
Accept	Accept	Accept

These different types of risk response plans for threats, opportunities, and overall project risk are expanded on and explained in the following sections.

7.6 Risk Response Plans for Negative Risks (Threats)

PMBOK recognizes five generic ways of treating risk for negative risks (i.e., threats). The risk response plan options are described below.

46 ISO 31000:2018 Section 6.4.4

7.6.1 Escalate

PMBOK 6th edition has introduced the concept of escalating risks.[47] If the threat is outside the scope of the project, or if the risk response plan would be beyond the PM's authority, then consider escalating the risk. Such threats should be managed at a level above the project (i.e., program, portfolio, or otherwise, as appropriate).

The PM should manage the notifications and communicate the escalation as appropriate. Obviously, people need to accept the risk if you escalate to them, so any escalation of risk should be a collaborative process. Record escalated threats in the risk register but don't manage them at the project level any longer.

7.6.2 Avoid

Evade risk completely by removing the cause of the risk or by changing the project planning or scope so that the risk no longer exists. This type of response plan is typically reserved for high priority risks, usually threats with both a high probability of occurring and a high impact.

Removing the threat from the scope of the project enables you to completely remove the probability of the risk occurring.

Examples of actions that project managers may take to avoid particular threats include

- Isolate the project objectives from the risk's impact
- Removing the cause of a threat
- Extending the schedule
- Changing the project strategy
- Reducing scope
- Clarifying requirements
- Obtaining more information
- Improving communications
- Acquiring expertise

7.6.3 Transfer

Transfer risk entails shifting the responsibility for the risk to a third party. The risk still exists, but the responsibility rests elsewhere. If the risk occurs, the impact will be felt by the third party.

Third parties who agree to manage the risk and bear the brunt of the impact if they occur, only do so at a price. There is a premium that the project will have to pay if they want to transfer the risk.

Examples of types of transfer of risk include but are not limited to

- Insure against the risk
- Issue or acquire performance bonds

[47] PMBOK Section 11.5.2.4 Page 442

- Negotiate warranties/guarantees and
- Enter into third-party agreements

It is rare that you can transfer all the inherent risk. When you use the transfer risk response, there is usually a residual and/or secondary risk that remains. If you have transferred a risk and you don't have a residual and/or secondary risk, it is worth revisiting the risk to check that you haven't missed a lingering uncertainty.

7.6.4 Mitigate

Define the action you are going to take to reduce the probability and/or reduce the impact of the risk being realized. Mitigating risks early in the project is often easier and less expensive than dealing with a risk after it has occurred. Mitigation is the most often employed risk response strategy in ICT projects.

Examples of actions that we can take to mitigate risks include but are not limited to

- Develop a prototype
- Target factors that increase the severity of the impact
- Target factors that increase the probability
- Designing redundancies into systems

7.6.5 Accept

Do nothing active to treat the risk. Take no immediate action on the risk; the risk is to remain on the risk register and include it in the monitor and review process.

There are two types of "accept" risk response plans: passive acceptance and active acceptance.

Passive acceptance of a risk requires no action beyond documenting the decision.

Active acceptance involves some further action. Actions could include, but are not limited to, one of the points below or any combination of the three.

- Setting aside some contingency reserves to offset the effect of the risk
- Producing a contingency plan
- Producing a fallback plan to enact if the risk is realized

7.7 Risk Response Plans for Positive Risks (Opportunities)

The risk response plans for opportunities are similar but distinct from the risk response plans for threats.

7.7.1 Escalate

PMBOK 6th edition has introduced the concept of escalating risks.[48]

[48] PMBOK Section 11.5.2.4 Page 442

Opportunities that are outside of the scope of the project and/or outside of the PM's authority can be escalated up to a level above the project. Escalate them to be managed at the program or portfolio level or to some other level specific to your organization.

Escalate them to the level that would be affected if the opportunity occurred. The PM manages the notifications and communicates the escalation as appropriate.

Obviously, the person you escalate the opportunity to needs to accept the opportunity, so any escalation should be a collaborative process. Record escalated opportunities in the risk register but don't manage them at the project level any longer.

7.7.2 Exploit

High priority opportunities that the organization wants to ensure that they reap the benefits of can be addressed with the exploit risk response plan. Exploiting is the equivalent/inverse of the threat response *avoid*. As *avoid* seeks to reduce the probability of a threat to 0%, *exploit* seeks to increase the probability of an opportunity to 100%.

Examples of exploit risk response plans include

- Assigning an organization's most talented resources to the project to help reduce the schedule
- Using new technologies to reduce costs and duration
- Using technology upgrades to reduce costs and duration

7.7.3 Share

Sharing involves transferring some of the opportunity risk to a third party. If the risk occurs, the positive risk will benefit both parties. Sharing opportunities is the equivalent/inverse of transfer for threats.

The sharing risk response plan can be a great way of improving your likelihood of achieving the associated benefits. This can be especially true if the third party that you are sharing with has something that you don't have. They might have the capability, capacity, knowledge, or resources that would enhance the probability of the opportunity occurring or that would improve the benefits that you could achieve through the opportunity.

Third parties are unlikely to take on uncertainty for free. They will want a risk premium that the project will have to pay if they want to share the opportunity. Or, they may insist on favorable terms of the sharing agreement such that they receive the lion's share of the benefits.

Examples of types of share risk response plans include but are not limited to

- Forming risk-sharing partnerships
- Creating special-purpose companies
- Entering into joint ventures

7.7.4 Enhance

Enhance entails defining the actions you are going to take to increase the probability and/or increase the impact of the opportunity being realized. Enhance for opportunities is the equivalent/inverse of mitigate for threats.

Examples of actions that we can take to mitigate risks include but are not limited to

- Assigning more qualified resources to the project
- Upgrading technologies to reduce costs and duration
- Developing a prototype
- Targeting factors that increase the severity of the impact
- Designing redundancies into systems

7.7.5 Accept

Take no active action. Take no immediate action regarding the opportunity. Leave the opportunity on the risk register and include it in the monitor and review process. This strategy might be the most appropriate when the opportunity is insignificant or if the cost/benefit ratio is not high enough to justify action.

Note: Passive acceptance of an opportunity requires no action beyond documenting the decision. Active acceptance involves some further action, either setting aside some contingency reserve of time, money and/or resources to take advantage of the opportunity should it be realized.

7.8 Risk Response Plans for Overall Project Risk

Risk responses to deal with overall project risk borrow from the strategies used to respond to individual project risks. If you have performed a quantitative risk analysis, then you will have the input and tools and techniques to create risk response plans for the overall project risk.

The risk response plans for dealing with an overall project risk are

7.8.1 Avoid

If the negative overall level of project risk is above the risk thresholds for the project, you may want to apply the avoid strategy.

The avoid strategy involves taking specific action to bring the overall project risk down below the risk threshold.

The avoid risk response plan might include options such as remove high-risk elements from the project scope. In extreme cases where project overall risk is persistently unacceptable, it may be necessary to cancel the project.

7.8.2 Exploit

If the level of overall project risk is high and very positive, an exploit strategy may be appropriate. An exploit strategy involves taking focused action to capture the positive risks on the project as a whole.

Examples of an exploit risk response plan for positive overall project risk might include, but are not limited to

- Adding high-benefit elements to the project scope to add value and benefit
- Modify the risk thresholds for the project to embrace the opportunity

Neither of those exploit strategies could be decided by the PM in isolation. The PM will need to liaise with the sponsor and/or board to determine if these risk response plans are appropriate and acceptable.

7.8.3 Transfer/Share

When the overall level of uncertainty for the project is high and the organization is unable to deal with it, it may be necessary to bring in a third party to transfer threats and/or share opportunities.

If the overall project risk is negative, then a transfer strategy may be most appropriate. Dealing with threats may involve pursuing any of the following risk response plans:

- Insure against the risk
- Issue or acquire performance bonds
- Negotiate warranties/guarantees
- Enter into third-party agreements

Conversely, if the overall project risk is positive, then you could share risk ownership to reap the associated benefits. Dealing with opportunities may involve pursuing any of the following risk response plans:

- Setting up collaborative business structures
- Entering joint ventures
- Launching a special purpose company
- Entering into third-party agreements

Any one of these options would help the project transfer/share the overall project risk to maximize the likelihood of achieving the project objectives.

7.8.4 Mitigate/Enhance

The mitigate/enhance strategies are used to improve the overall project risk to maximize the project's chances of achieving its stated objectives.

Mitigate involves defining the action you will take to reduce the probability and/or reduce the impact of a negative overall project risk being realized.

Enhance entails defining the action you will take to increase the probability and/or increase the impact of a positive overall project risk being realized.

Examples of actions that we can take to mitigate/enhance overall project risks include but are not limited to

- Changing scheduled delivery
- Allocating resources differently
- Amending the scope of the project
- Re-planning the project
- Changing the project priorities

7.8.5 Accept

If no risk response plan could be practically implemented for an overall project risk, the organization may decide to accept the risk. This strategy might be the most appropriate when the risk is insignificant or if the cost/benefit ratio is not high enough to justify action.

Note: Passive acceptance of an overall project risk requires no action beyond documenting the decision. Active acceptance involves some further action, usually setting aside some contingency reserve of time, money and/or resources to address the risk should it be realized.

7.9 Contingent Risk Response Plans

Contingent risk response plans are part of the *active accept* risk response plan, but they have some complexity that justifies describing them separately.

Contingent risk response plans are plans that you develop as part of the plan risk responses process, but you do not act on until an event triggers them. We prepare these plans up front, but we don't them put into action until an event triggers them. And if no event triggers them, they are never enacted.

Contingent response plans are risk response plans that we develop for use only if a risk is realized. For some circumstances, the most appropriate response plan may be an action that we will only take in reaction to a risk trigger as it occurs. If such reactions and their triggering events are defined in advance, they are considered contingent response plans.

The nature of some risks is such that a contingency plan is the most appropriate way to plan to respond to the risk. If you believe that the triggering event will provide sufficient time to enact the predefined contingency plan, then a contingency plan may be the most efficient way of responding to the risk. It often costs less, and it can be just as effective to plan a contingency plan than it would be to implement a mitigation plan.

If you do use contingency plans, you will need to have a mechanism to identify and monitor risk triggers. And, you will need to have a mechanism to ensure that the how, what, where, when, and whom details are kept up to date for the contingency plan.

If you use contingent response plans, you will need to have a mechanism to measure and track the triggering conditions. And, you will need a mechanism to monitor the contingent response plans to ensure that they are still the most appropriate responses. You will also need a mechanism to ensure that all the required resources are still in place to react if the risk is triggered.[49]

7.10 Integrated Risk Management – Escalation and Receiving Delegation

Risks exist at all levels of an organization. Risks may be identified at a level that is not appropriate for them to be managed at. If they are identified at a level other than the level where the impact would be felt, then they need to be reassigned to the appropriate level.

If the risk response planning process reveals that risks are at the incorrect level, they should be escalated or delegated as appropriate.

High-level risks that have you have identified at the project level should be escalated to the program, portfolio, or strategic level.

Conversely, project related risks that have been identified as the program, portfolio, or strategic level should be delegated to the project. The project should have a pre-defined mechanism for receiving and integrating any risks that are delegated down from the program, portfolio, or strategic level. Ideally, you will have defined this escalation and delegation mechanism in the risk management plan.

This is depicted in the Escalation & Delegation Diagram below. The yellow arrows indicate newly discovered risks. The green rectangles represent the level they are discovered at. The reddish arrows represent escalation of risks. The blue arrows represent the delegation of risks.

Diagram 36 – Escalation & Delegation Diagram

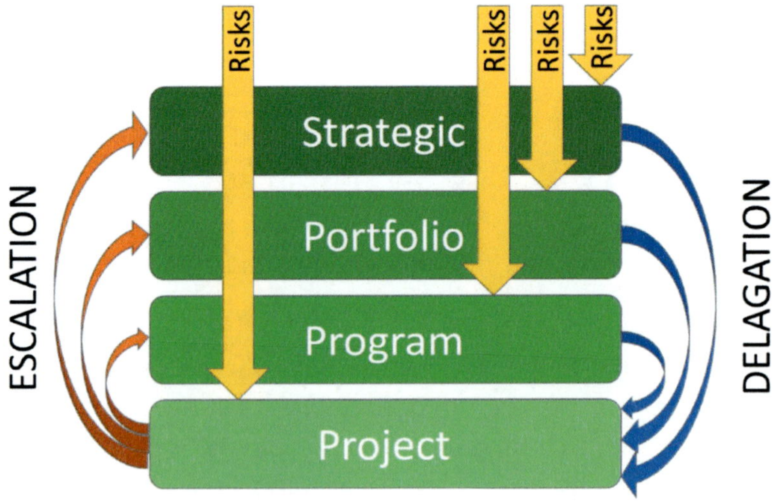

[49] PMBOK Section 11.5

7.11 Risk Response Plans Content

The information provided in the risk response plan should be complete enough for the relevant people to be able to execute the response plan. A risk response plan should include

- The rationale for the selection of the treatment options, including the expected benefits
- The names and/or positions of those who are accountable and responsible for approving and implementing the plan
- The proposed actions
- The resources required, including contingency reserves
- The performance measures
- The constraints
- The reporting and monitoring that is required
- When actions are expected to be started and completed[50]

We may be limited in the risk response information that we can capture in the risk register, but this doesn't need to be the only place you capture risk response information. I would suggest that for complicated/complex risks you should have a secondary place where you store the additional risk response information.

Having risk response information stored in two separate locations adds complexity and will require a mechanism to manage the information to keep it in sync. If you do keep risk response information outside of the risk register, you will need to define how, when, where, how often, and by whom the information is updated and kept in sync.

7.12 Risk Levels

As described in section 4.3.9, in the workshop you tried to establish the most appropriate level of each risk. Any risks that have slipped through and are still identified at an incorrect level will need to be either rolled up or split up, so they are described at the most appropriate level before a risk response plan can be devised.

When assessing risk level, we're looking for the Goldilocks zone of risk description. If risks are at too high a level, it is difficult to apply a meaningful and actionable risk response plan. Conversely, if risks have been described at too low a level, the risks may individually be assigned low risk scores and, as a consequence, slide under the risk threshold. We need to find the level of risk description that is "just right."

[50] ISO 31000:2018 Section 6.5.3

There is a danger that several granular risks are listed individually that could have been addressed through the same risk response plan. In such instances, these risks slide under the risk threshold and the risk response plan "accept risk" is often applied.

For example, let's say you have identified a risk for finance staff availability, you have identified another risk for operations staff availability, and you have identified a third risk for IT staff availability. If you list these risks separately, each risk may not justify active treatment. They may be passively accepted. Whereas, if you roll them up into one risk—perhaps "staff availability"—the risk may exceed a risk threshold and you would apply an active risk response plan (e.g., mitigate).

By this point in the risk management process, you have already considered the level of the individual risks twice. This is the final opportunity to check to ensure that you have recorded the risks at the most appropriate level.

7.13 Cost/Benefit Analysis

As alluded to in section 6.2, a cost/benefit analysis is required to ensure the risk response plan choice is appropriate.

The straightforward application of a risk threshold does not take into consideration the cost/benefit of each risk response plan. You may have a risk that falls under a risk threshold but can be treated (mitigated) with very little time and effort (e.g., by a short telephone call).

Conversely, you may have a risk that is over the risk threshold but will require significant time and effort to mitigate. In such a situation, it may be more sensible to apply project resources (e.g., mitigate) to the first risk than it would be for the second risk, even though the first risk is a lower risk score.

Similarly, a cost/benefit analysis should be applied when choosing the risk response plan that you will apply to the project risk. Generally speaking, it doesn't make much sense to spend more time/resources on mitigating a risk than the expected monetary value (EMV) of a risk.

For example, if you have a risk that could cost you $100,000 and has a probability of occurring of 20%, it has an EMV of $20,000 (100,000 x 20%). In such a case, you typically wouldn't spend more than $20,000 in treating that risk.

Such a calculation is easy for financial benefits, but with some effort, we can also usually apply financial values to non-financial benefits.

You will have to assess your project's circumstances, but as a general rule, you would not typically spend more treating a risk than the maximum potential cost of the impact.

Depending on your sponsor's risk appetite and the nature of your project, there may be situations where you would spend more than the EMV of a risk to treat a risk.

For example, if your project sponsor has a budget that is high, but a schedule that absolutely cannot be exceeded, it may make sense to exceed a risk's EMV in treating schedule risks.

If you have a project that has a hard deadline that it absolutely "must meet," you may spend excessive resources, well in excess of a risk's EMV, to treat risks. If the nature of your project dictates it, swallowing inefficiencies in treating risks to ensure you hit a hard deadline may make sense.

For example, if you're delivering an IT system to support an event like the Olympic Games, the deadline is hard and unmovable. You would bear inefficiencies in treating risks to ensure that your system is delivered on time. If delivering a month late would render the whole project useless, it would make sense to spend above the EMV on schedule risks to ensure you hit the deadline.

If you find yourself in a situation where you are planning to spend more treating a risk than the impact of the risk could potentially cost, you might want to check that you have calculated the potential impact correctly.

I remember observing a risk review meeting where a PM stated that he wanted to mitigate any risk that could be mitigated. If he did that without any consideration of cost/benefit analysis he could have easily doubled the cost of the project!

Assess your project sponsor's risk appetite, assess how much reserves you have available, and assess how hard your project deadlines are. Then, use a cost/benefit analysis to determine the risk response plans that are most appropriate for your project.

7.14 Decision-Making for Risk Response Plans

There are tools and techniques that we can use to help us make decisions on risk response plans. We can use these decision-making tools and techniques to help select and prioritize risk response plans.

The tools and techniques that we can use to help with decision making include but are not limited to: multi-criteria decision analysis (MCDA), analytical hierarchy process (AHP), decision trees, and influence diagrams. How you can use these tools and techniques is described in more detail in sections 3.2 and 6.5.

7.14.1 Risk Threshold

Due to the potential granular nature of some risks and cost/benefit variances amongst risk response plans, I would urge caution in the straightforward application of a risk threshold when evaluating risk response plans. A risk threshold is an objective tool, and I recommend that you also apply a subjective assessment.

Subjectively consider any risks that fall below the threshold. Are they described at the most appropriate level? Could you combine them with other risks? Are there treatments for these risks that constitute "low-hanging fruit"? Are there simple and quick actions that can treat these risks? Is the cost of a risk response plan so low that it's worth doing despite the risk passing under the risk threshold? If so, consider applying a risk response plan.

7.14.2 Risk Response Plan Process

When creating the risk response plans, it is important to consider each of the following:

- Options - what risk response plans are possible for each risk?
- Residual risk – how much of the risk does your risk response plan actually address?
- Secondary risk – does the risk response plan itself introduce any secondary risk?
- Cost/benefit analysis – how much is it going to cost to achieve the benefit by treating the risk? How big is the benefit? What is the cost/benefit for each of the risk response plan options?
- Effort – how much effort will you need for each risk response plan option? Is it realistic and achievable given the project constraints?
- Ownership and execution – who is going to own, manage, and execute the risk response plan? Does your project have the capability and capacity?

When developing risk response plans, work through the steps outlined, below. The steps in the develop risk response plan process are

1. Review risk against threshold – review each risk against the risk threshold to identify which ones are above and which ones are below the threshold
2. Options – determine which risk response plans are possible for each risk
3. Question cost vs. benefit – assess the value of each proposed risk response plan options against the time and cost of implementing it; assess the risks below the risk threshold as well to see if there are any low-hanging fruit
4. Determine the risk response plans – decide the risk response plans that are most appropriate for the given risk appetite, resources available, and value
5. Update risk register – circulate the risk response plans to the stakeholders; update the risk register with “proposed” status risks, update with participants’ feedback, and amend the status; assign the risk to someone, then record and track the tasks

7.14.3 Manage the Time You Spend on Planning

The time you spend on risk response planning should be commensurate with the level of risk and the importance of the project. Avoid the temptation to spend time trying to create the perfect plan.

> *“A good plan today is better than a perfect plan tomorrow.”*
>
> George Patton – US General

It would be easy to lose yourself for days in planning risk responses. Days of planning may be appropriate for significant risks on huge, existential projects. But, it may be overkill for medium-sized projects or projects that are not critical for the organization’s existence. It is good practice to keep the risk response plans appropriate to the size and importance of your project. Keep the risk response plans as simple as possible.

"Everything should be made as simple as possible, but no simpler."

Albert Einstein, German-born theoretical physicist

ICT projects tend to be fast moving environments. In project risk management we need to find a balance between being thorough and being timely. We need to work our way through the risk management process to the appropriate level of detail. But, we also need to deliver results as quickly as possible so that the risk response plans remain timely and relevant.

There is no hard and fast rule on how much time you should spend on risk response planning. By now, you will have a good understanding of your project's importance, your sponsor's risk appetite, your project's overall risk, and the number and severity of your individual project risks. You'll need to use that knowledge to determine how much time is appropriate to spend on risk response planning for your project.

7.14.4 Iterative Quantitative Analysis and Risk Response Plans

Many of your project's individual risks will have changed after you've decided on your risk response plans. You will have used expert judgement to assess the residual / secondary risk levels of individual risks as part of the risk response plan process.

Overall risk exposure is also almost certain to have changed after you determine your risk response plans. But, the impact of the risk response plans on overall project risk cannot be quantified without rerunning the quantitative risk analysis.

Rerunning the quantitative risk analysis Monte Carlo simulations enables us to determine how our risk response plans have affected the overall project risk exposure. It will also enable us to assess which individual risks most impact the project after the risk response plans have been enacted.

Rerunning the quantitative risk analysis Monte Carlo simulations after having decided on the risk response plans will enable us to find the new answers to our four very important questions:

1. What is the new probability of achieving at least our stated ROI or scheduled completion date given the risk response plans that we have implemented?
2. What is the new probability of losing money on the project or missing a delivery deadline given the risk response plans that we have implemented?
3. What is the new contingency reserve we would need to be 80% (or another percentage) sure of achieving the stated cost or schedule objectives given the risk response plans that we have implemented?
4. What are the variables that have the largest effect on the probability of the project's outcome given the risk response plans that we have implemented?

The answers to those questions are almost certain to have changed after enacting the risk response plans.

Your project sponsor may find the answers to those questions are still unacceptable. If that is the case, you will need to increase the amount of time and effort to further reduce your overall project risk exposure.

You may need to keep repeating this cycle until your project sponsor is content with the overall project risk exposure that you achieve.

If your project is large, important, or complex, you will want to perform a quantitative risk analysis following the plan risk responses process to determine the probable success of planned responses in reducing overall project risk exposure.[51]

7.15 Risk Response Plans and Risk Thresholds

The risk management methodologies implemented in organizations often stipulates the application of a risk threshold based on the risk score. The use of a risk threshold allows for risks with a risk score below a pre-established value to be tolerated as they fall within the organization's or project's risk appetite.

Although that can be used as a simple and neat method to exclude some risks from the risk response plan process (and thereby reduce the amount of effort required), I would exercise caution when simply disregarding risks below any threshold. There are two reasons for this.

Firstly, establishing the level of the risk is prone to error, and blindly disregarding risks below a threshold prevents you from performing a "sanity check" on the risk level. It is worth running through the risks below the threshold to ensure that they aren't risks that should be rolled up with other risks, which might cause them to exceed the threshold.

Secondly, the application of a risk threshold does not account for the varying effort, cost, and benefit for risk response plans of different risks. There may be risks below the threshold that could be treated with little time and effort. It is worth running through the risks below the threshold to see if there are any "quick wins."

7.16 Conclusion of the Workshop

At the conclusion of your workshop, it is good practice to set out the next steps. Stakeholders who have participated in the workshop will be curious to know how you will use their contribution. They deserve to know that the time that they have put in will be of benefit and that you value their participation.

Once you have written up the risks, it is best practice to circulate the output of the workshop to the participants for review. We do this both to ask for feedback on any errors/omissions and to gain tacit approval.

Now that you have established a foundation of risks and risk response plans, you'll need to update the information in the risk register so that it can be accessed and managed.

[51] PMBOK 11.4 page 429

Risk response plans that will be implemented will need to be captured in the project plan and other project documentation as tasks to be tracked and reported on. This may require the risk to be tracked in several places: the project plan and the risk register. You will need a mechanism for keeping any updates in sync if there are several places where the risk is recorded.

7.17 Chapter Summary – Plan Risk Response

Risk response plans define the approach that a project takes when dealing with identified risks. The first option that springs to mind for most people is to mitigate the risk. Indeed, this is often the approach taken. But, depending on your risk attitude and the cost of mitigation, you may choose to go down a different path. You have several different options to treat overall risk, threats, and opportunities. Or, if the cost of these risk response plans is too high, you may choose to accept the risk.

In this chapter we have covered

- The plan risk response process
- The risk response plan options for threats including escalate, avoid, transfer, mitigate, and accept.
- The risk response plan options for opportunities including escalate, exploit, share, enhance, and accept
- The risk response plan options for overall project risk including avoid, exploit, transfer/share, mitigate/enhance, and accept
- Contingent responses and how they augment actively accepted risk response plans
- Integrated risk management including the delegation and escalation of risks and how they integrate into broader project management
- The content of a risk response plan and how the detail can be managed
- How risk levels can impact risk response plan decisions and how to manage them
- How cost/benefit analysis can be applied to help determine the appropriate risk response plans
- How decision making for the risk response plans can be handled including risk threshold, risk response plan process, time management, and iterative quantitative analysis

And finally, we covered the conclusion of the risk workshop for those who have included risk response plans in their risk workshop.

8 Implement Risk Response Plans

Implement risk response plans is the sixth link in our risk management process chain of steps.

Diagram 37 – Implement Risk Response Plans

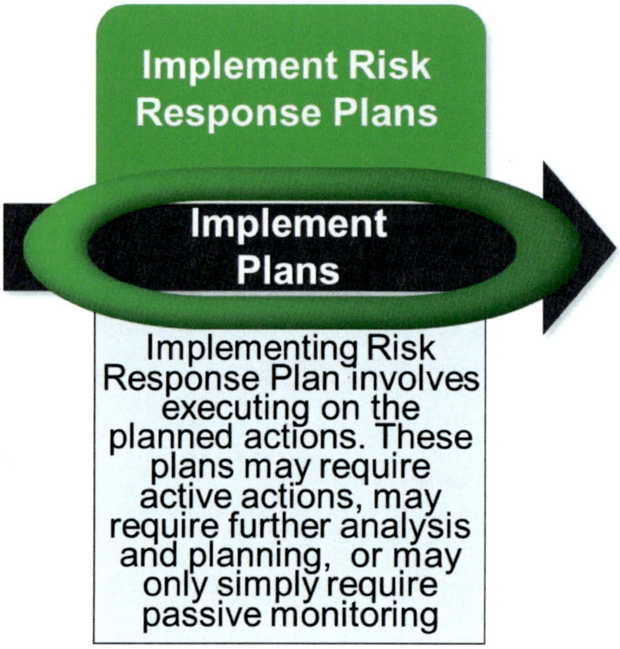

Once you have determined your risk response plans, you'll need to implement them if you're going to manage your project's risk exposure as intended.

You can start this process as soon as you complete the first risk response plan, and you will continue until you close the project.

Some of the risk response plans will be easier to implement than others. And, some of them you will be able to defer implementing. But you should track all of them in the risk register. And you should integrate all of them into any other relevant project documentation.

One of the most important documents to update is the project schedule. This is the project document that is going to be the most helpful in implementing the risk response plans.

If you're going to reap the benefits of the risk response plans, you need to put a process in place to ensure that the risk response plans are executed as intended. Your process should ensure that expectations are communicated and that there is a method to monitor, control, and track risk response plans. And, your process should include a mechanism for addressing any risk response plans that don't achieve the intended results.

8.1 Implementing Risk Response Plans for Threats

8.1.1 Escalate

If the risk response plan for a threat has been identified as being outside of the scope of the project, or if the response would be outside of the project manager's authority, then you should *escalate* the risk.

Assess the risk and consult the risk management plan. The plan should instruct you on how to escalate the risk, who you should escalate the risk to and when you should escalate the risk.

Escalation should be collaborative to ensure that the risk is accepted by the party that you are escalating it to. If there is any resistance or conflict, seek guidance from the project sponsor and/or the project board.

Update the risk register with the actions taken and update any related project documentation. There is no need to further monitor the risk.

8.1.2 Avoid

If you have identified a risk that has a high probability and a large negative impact, you may decide to *avoid* it completely. Avoiding reduces the probability of it occurring down to zero percent.

Depending on the source of the risk, you could use different methods to avoid the risk.

You might be able to avoid the risk by extending the schedule. This might be especially effective for a risk whose source pertains to holiday periods or seasonal events.

If the source of a risk relates to the project strategy, you might be able to avoid the risk by changing the project strategy.

If the source of the risk is a particular element, function, or interface of the project, you may be able to avoid the risk by descoping the project. If there is a particular element, function, or interface that is a source of great risk, you could simply exclude it from the scope.

Obviously, these are all large changes and would almost certainly need to go through your change control process. Your risk management plan should describe how the change control process works for your project.

Update the risk register with the actions taken. And, update any other relevant project documentation. There is no need to further monitor the risk.

8.1.3 Transfer

If the nature of a risk is such that a transfer risk response plan is feasible, you may choose to *transfer* the risk.

You can transfer a risk through a number of methods. You may be able to take out an insurance policy against the risk. You might be able to arrange for performance bonds, warranties, guarantees, or any other agreement that can transfer ownership of the risk to another party.

Transferring risk rarely completely eradicates the risk from your project. Transferring usually leaves you with some residual risk and/or secondary risk.

For example, insurance contracts often have a deductible/excess clause where your contract will have to bear the risk for the first percentage or defined amount. Such a clause would leave your project with a secondary risk for the amount of the deductible/excess.

Almost every other method of transferring risk will result in your project wearing at least some counterparty risk. What happens if the party that you've agreed to transfer the risk to goes bankrupt? That is one example of a secondary risk that your project may still be exposed to.

Update the risk register with the actions taken and record any residual and/or secondary risks. You may decide to create risk response plans for both the residual and secondary risks, and you will need to continue to monitor both the residual and secondary risks and update any other relevant project documentation.

8.1.4 Mitigate

If a risk *mitigation* option is cost effective, you may determine that risk mitigation is the most appropriate risk response plan.

Mitigating a risk means that you either reduce the probability of the risk being realized and/or you reduce the impact that risk would have on the project if the risk was realized.

If you have determined that mitigation is the most appropriate risk response plan for a risk, you will need to identify all the actions that you will need to take to mitigate the risk.

Once you have mapped out all the actions, you will need to integrate those tasks with the other project documentation.

For example, you'll need to integrate the risk mitigation actions into your schedule and manage any dependencies. You'll also need to ensure that you adjust your resource plan to accommodate the extra activities. And, you'll need to adjust your budget to ensure all the required activities are funded.

Once you have updated all the project documentation with the mitigation plan, you'll need to monitor the mitigations process and keep the risk register up to date.

8.1.5 Accept

If a risk is low priority, or if you can't treat it cost-effectively in any other way, then you may determine that *accepting* the risk is the most appropriate risk response plan.

The accept risk response plan has two separate types of acceptance. We divide "accept" into passive acceptance and active acceptance.

Passive acceptance simply involves updating the risk register and taking no further action.

Active acceptance is more complicated. You still need to update the risk register, but you also may have to arrange for a contingency reserve in case the risk is realized.

You may also determine that a contingency plan and/or fallback plan are appropriate. Contingency and fallback plans are not enacted until a pre-defined event triggers them, but you need to prepare them at the outset, and you should make funding arrangements.

Any active acceptance risk response plan will likely also require arrangement of extra funding and updating of the budget.

Update the risk register with any actions taken. You will need to continue to monitor the risk triggers for any contingency plans, and you will need to regularly review contingency plans to keep them up to date. You will also need to update any other relevant project documentation.

8.2 Implementing Risk Response Plans for Opportunities

8.2.1 Escalate

If the risk response plan for an opportunity has been identified as being outside of the scope of the project, or if the response would be outside of the project manager's authority, then you should *escalate* the risk.

Assess the risk and consult the risk management plan. The plan should instruct you on how to escalate the risk, who you should escalate the risk to and when you should escalate the risk.

Ensure that the party that you are escalating to accepts the risk. If there is any resistance or conflict, seek guidance from the project sponsor and/or the project board.

Update the risk register with the actions taken, but there is no need to further monitor the risk and update any other relevant project documentation.

8.2.2 Exploit

High-priority opportunities may justify the *exploit* risk response plan. If the organization is especially keen to ensure they realize an opportunity, they will choose to exploit the opportunity.

Exploiting an opportunity will require the project to actively change tasks, resources, and/or budgets to ensure they realize the opportunity. And you will need to record any changes you make to the project in the relevant project documentation. You need to monitor and manage any changes to ensure that you realize the opportunities.

You should update the risk register with the approach, and you should regularly review the approach.

8.2.3 Share

Projects typically decide to *share* opportunities when they don't have the capability or capacity to enhance or exploit the opportunity by themselves.

Sharing an opportunity may take many potential forms including, but not limited to, a joint venture, a new team or a new partnership.

Sharing an opportunity may involve giving a risk premium to a third party. This may be a payment, or it may be an unequal sharing of the benefits on terms that are favorable to the third party. You will need to agree and document any contractual and funding arrangements.

Whatever the risk premium, the synergies of dealing with a third party that has the capability and/or capacity to help realize the opportunity should result in your organization reaping rewards beyond what you would have been able to on your own.

Sharing often requires legal agreements that require input from legal and procurement departments. You will need to document any of these agreements.

Sharing an opportunity requires an update of relevant project documentation. You will need to monitor and document all the extra tasks, and you will need a mechanism to keep them up to date.

Update the risk register with the approach and review the risk regularly.

8.2.4 Enhance

If an opportunity is appealing but not high-priority, you may determine that *enhance* risk is the most appropriate risk response plan.

Enhancing a risk means that you either increase the probability of realizing the risk and/or you increase the impact that risk would have on the project if the risk was realized.

If you have determined that enhancement is the most appropriate risk response plan for a risk, you will need to identify all the actions that need to be taken to enhance the risk.

Once you have mapped out all the tasks you need to take, you need to integrate those tasks with the other project documentation. For example, you'll need to integrate the tasks into your schedule and manage any dependencies. You'll need to adjust your resource plan to accommodate the additional activities, and you'll need to adjust your budget to ensure all the required activities are funded.

Once you have updated all the project documentation with the risk response plan, you'll need to monitor the enhancements and keep the risk register up to date.

8.2.5 Accept

If you can't enhance or exploit an opportunity cost effectively, or if it is low priority, then you may determine that *accepting* the risk is the most appropriate risk response plan.

The accept risk response plan actually has two separate types. Accept can be divided into passive acceptance and active acceptance.

Passive acceptance simply involves updating the risk register and taking no further action other than a periodic review.

Active acceptance is more complicated. You still need to update the risk register, but you may also have to arrange for a contingency reserve to hold in case the risk is realized, as you may need to fund actions to realize the benefits of the opportunity.

You may also determine that a contingency plan is appropriate. Contingency plans are not enacted until a pre-defined event triggers them, but you need to prepare them at the outset.

Any active acceptance risk response plan will likely also require negotiation of additional funding arrangements for contingency tasks and updating of the budget.

Update the risk register with any actions taken. You will need to continue to monitor the risk triggers for any contingency plans, and you will need to ensure contingency plans are regularly reviewed and kept up to date. And, you will need to update any other relevant project documentation.

8.3 Implementing Risk Response Plans for Overall Project Risk

8.3.1 Avoid

If you have assessed the overall project risk as highly negative, you may determine that the *avoid* risk response plan is most appropriate. We use the avoid risk response plan when the negative overall project risk is above the project risk threshold.

The avoid risk response plan for overall project risks requires that you engage in actions to reduce the negative effects of uncertainty on the project as a whole.

If there are elements of the project that contribute significantly to the overall project risk, you may be able to avoid the risk by descoping the project. If there is a particular element, function, or interface that is a source of great risk, you could simply exclude it from the scope.

The sensitivity analysis performed in the Monte Carlo simulations as part of quantitative analysis will help to identify the elements of the project that contribute the most to negative uncertainty.

Update the risk register with the actions taken and update any other relevant project documentation.

In extreme cases, if you can't bring the overall project risk exposure back within the risk thresholds, it may be necessary to cancel the project. Such an extreme event would trigger the project closure process.

8.3.2 Exploit

High-priority overall project opportunities that exceed the remit of the project may justify the *exploit overall project* risk response plan. If the organization is especially keen to ensure that they realize an opportunity, they may choose to exploit the opportunity.

Exploiting an overall project risk will require the project to actively make changes to activities, resources, and/or budgets. These changes will likely be large enough that they need to be approved through the change control process. Your risk management plan should provide you with guidance.

And, you will need to update any changes made in the relevant project documentation. You will need to monitor and manage any changes to ensure that you realize the opportunities.

You should update the risk register with the approach, and you should regularly review the risk.

8.3.3 Transfer/Share

If the overall project risk is high but the organization does not have the capacity or capability to manage the risk on their own, they may look to a third party to help them manage the risk.

If the nature of a negative overall project risk is such that you could enact a *transfer* risk response plan, you may choose to transfer the risk.

You can transfer a negative overall project risk through several methods. You may be able to take out an insurance policy against the risk. You might be able to arrange for performance bonds, warranties, guarantees, or any other agreement that can transfer ownership of the risk to another party.

As with transferring individual threats, transferring negative overall project risk rarely completely eradicates the risk from your project. Transferring usually leaves you with residual risk and/or secondary risk. Any residual risk and/or secondary risk will need to be considered for contingency reserve funding. And, they may merit risk response plans of their own and will need to be tracked.

Sharing a positive overall project risk may take many potential forms including, but not limited to, a joint venture, a new team, or a new partnership.

Sharing a positive overall project risk may involve giving a risk premium to the third party. This may be a payment, or it may be a sharing of the benefits that is favorable to the third party. Any payment or benefit sharing will likely necessitate the creation or updating of contractual and funding documentation.

Whatever the risk premium, the synergies of dealing with a third party that has the capability and/or capacity to help realize the opportunity should result in your organization reaping rewards beyond what you would have been able to on their own.

Sharing often requires agreements that require input from legal and procurement departments, which you will need to arrange.

Regardless of whether you are transferring a negative overall project risk or if you are sharing a positive overall project risk, both require an update of relevant project documentation. And, you will need to monitor and keep all the additional tasks up to date.

You should update the risk register with the approach, and you should regularly review the risk.

8.3.4 Mitigate/Enhance

You may want to change your project's exposure to an overall project risk. If it is a negative overall project risk, you want to *mitigate* the risk. If it is a positive overall project risk, you want to *enhance* the risk.

If a negative overall project risk mitigation option is cost effective, you may determine that risk mitigation is the most appropriate risk response plan.

Mitigating a risk means that you either reduce the probability of realizing the risk and/or you reduce the impact that risk would have on the project if the risk were realized.

If you have determined that mitigation is the most appropriate risk response plan for a negative overall project risk, you will need to identify all the actions that need to be taken to mitigate the risk.

Once you have mapped out all the actions, you will need to integrate those tasks with the other project documentation.

For example, you'll need to integrate the risk mitigation actions into your schedule and manage any dependencies. You'll also need to adjust your resource plan to accommodate the additional activities. And, you'll need to adjust your budget to ensure all the required activities are funded.

Once you have updated all the project documentation with the mitigation plan, you'll need to track the mitigations process and keep the risk register up to date.

If a positive overall project risk is appealing but not high-priority, you may determine that risk enhancement is the most appropriate risk response plan.

Enhancing a risk means that you either increase the probability of the risk being realized and/or you increase the impact that risk would have on the project if the risk was realized.

If you have determined that enhancement is the most appropriate risk response plan for a positive overall project risk, you will need to identify all the actions that need to be taken to enhance the risk.

Once you have mapped out all the actions, you will need to integrate those tasks with the other project documentation. For example, you'll need to integrate the actions into your schedule and manage any dependencies. You'll need to ensure that you adjust your resource plan to accommodate the additional activities. And, you'll need to amend your budget to ensure all the required activities are funded.

Once you have updated all the project documentation with the risk response plan, you'll need to monitor the mitigations and keep the risk register up to date.

8.3.5 Accept

If an overall project risk is low priority, or if the risk can't be treated cost-effectively, then you may determine that *accepting* the risk is the most appropriate risk response plan.

The accept risk response plan has two separate types. Accept can be divided into *passive* acceptance and *active* acceptance.

Passive acceptance simply involves updating the risk register and taking no further action.

Active acceptance is more complicated. You still need to update the risk register, but you may also have to arrange for a contingency reserve to hold in case the risk is realized, as you need to fund actions to treat the overall project risk.

You may also determine that a contingency plan is appropriate. Contingency plans are not enacted until a pre-defined event triggers them, but you need to prepare them at the outset.

Any active acceptance risk response plan will likely also require negotiation of additional funding arrangements for contingency tasks and updating of the budget.

Update the risk register with any actions taken. You will need to continue to monitor the risk triggers for any contingency plans, and you will need to ensure contingency plans are regularly reviewed for relevance and kept up to date. You should also update any other relevant project documentation.

8.4 Chapter Summary – Implement Risk Response Plans

The actions required to implement risk response plans vary according to whether the risk is a threat, an opportunity, or an overall project risk. The risk response plans may include but are not limited to

- Implement risk response plans for *threats* including
  - Escalate
  - Avoid
  - Transfer
  - Mitigate
  - Accept
- Implement risk response plans for *opportunities* including
  - Escalate
  - Exploit
  - Share
  - Enhance
  - Accept
- Implement risk response plans for *overall project risk* including
  - Avoid
  - Exploit
  - Transfer/share
  - Mitigate/Enhance
  - Accept

Overall project risk response plans may necessitate that you raise a change request if the change is outside of your pre-determined tolerances.

You should record any actions you take to implement risk response plans in the appropriate project documentation.

9 Monitor and Control Risks

Monitor and control is the seventh and final link in our risk management process chain of steps.

Diagram 38 – Monitor and Control

Risk management plans on their own cannot ensure project success. The performance of the project needs to be continuously measured against the plan. You'll need to make assessments and corrections as you travel through the project lifecycle.

The monitor and control process is the process of scrutinizing the implementation of both the overall project risk and individual risk response plans. The process includes the tracking of identified risks, the identifying and analyzing of emergent risks, and evaluating the risk process effectiveness.

The monitor and control process enables project decisions to be based on current information on both overall project risk exposure and individual project risks. This process is performed iteratively throughout the project. Although the risk management process is frequently described as sequential, in practice it is iterative.[52]

You should continuously monitor your project for individual project risks that are new, have changed, and/or have become outdated.

[52] ISO 31000:2018 Section 6.1

You should also monitor overall project risk to ensure that your project stays within the sponsor's risk thresholds.

The risk response plans for both overall project risk and individual risks should be continuously monitored to ensure that they remain effective and appropriate.

You can reference the details of the monitor and control process in the risk management plan. Those details will include the *what*, *why*, *where*, *how*, *who*, *when*, and *how often* of the monitor and control process.

The monitor and control process utilizes both objective and subjective inputs. Technical performance analysis (objective), reserve analysis (objective), and expert judgement (subjective) will feed into the monitor and control process.

That information may be recorded in any number of project documents including, but not limited to, the issue log, the lessons learned register, the risk register, and the risk report.

There may also be pertinent information that has not yet been documented but that should be an input into the monitor and control process.

The monitor risks process uses performance information to determine if

- Implemented risk responses have been effective
- Level of overall project risk has changed
- Status of identified individual project risks has changed
- New individual project risks have arisen
- The risk management approach is still appropriate
- Risk management policies and procedures are being followed
- Contingency reserves for cost or schedule need modification
- Management reserves are still adequate
- Project risk strategy is still valid
- Project assumptions are still valid

ICT projects are dynamic, and so are their risks. The monitor and control process allows you to keep your finger on the pulse of the project's risks. The process keeps you abreast of any changes related to risks and enables you to respond to any changes early and swiftly. The monitor and control process is critical to successfully meeting the project objectives.

9.1 Identify and Analyze Emergent Risks

ICT projects are complex and complicated. As such, it is rare that those running ICT projects manage at the outset to identify all the risks that their projects will face.

On ICT projects, our initial identification of individual risks is usually only a first attempt. We recognize that, at the outset, we are not going to be able to identify every possible risk in our first effort. Recognizing this limitation encourages us to plan tasks to reconsider risk throughout the project, repeatedly searching for new emergent risks.

Anytime we identify emergent risks throughout the project lifecycle, we will put them through the same processes that we put our original risks through.

When we identify and evaluate emergent risk, we can reference the risk management plan which will describe the who, how, when, why and how often we intend to formally identify and analyze emergent risks.

9.2 Tracking Identified Individual Risks

Once you have entered a risk onto the risk register, you should track it throughout the project until the risk is "closed." You may decide to review different priority risks at different frequencies.

You might review all the high-priority risks (i.e., those on your *urgent* list) more frequently than you review the low-priority risks (i.e., those on your *watch list*). For example, you might review risks on your urgent list once per week, but you might only review risks on your watch list once per month.

Regardless of whether your risks are on your urgent list or on your watch list, you should review them regularly to ensure that your project doesn't encounter any surprises. You should review the risks to ensure that they are still relevant, that their impact and probability values have not changed, and that nothing else has changed about them.

If anything has changed on the risks in your risk register, you need to assess the effect of that change. If the change has had a material impact on your project and/or exceeded risk tolerances, you may need to escalate to determine your next steps.

If the change is immaterial, or is within your risk tolerances, at the very least you will need to update the risk register. And you will need to ensure that the risk owner and/or risk action taker are aware of the change, any actions required, and its implication.

9.3 Monitor the Implementation of Risk Response Plans

The risk response plans that we set in place for projects risks are devised as the most appropriate plans at the time. But, as with any plan, the realization of the risk response plan may be less successful than we had originally hoped.

As the project progresses, factors may change that render our risk response plans less appropriate. Internal factors may change that limit the success of the risk response plans. External changes might reduce the appropriateness or effectiveness of our risk response plans. A myriad of changes might occur that could impact the success of our best-laid risk response plans.

> *"The best-laid plans of mice and men often go awry."*
>
> Robert Burns, Scottish poet and lyricist

When we acknowledge that some of our plans will likely go awry over time, we are more inclined to recognize the need to monitor the effectiveness of those plans. And we are more prepared to react when some of those plans fail to achieve the results that we had hoped.

In the risk management plan, we outlined how, when, and how often we were going to monitor risk response plans. We should execute the monitoring of risk response plans in the method and to the timing outlined in the risk management plan.

We should periodically review each of our risk response plans to check to see if they have achieved the desired results and/or if they are still the most appropriate approach.

If we have risk response plans that have been executed, we should check to see if they have achieved the desired results. We should check to see if, after enacting the risk plan, the residual or secondary risk has been reduced as much as we had anticipated. We should update any discrepancy in the risk register and relevant project documents

If the actual residual/secondary risk has not achieved targeted value, we need to decide if the actual residual/secondary risk is still within risk tolerances. If it is not, we need to determine what we are going to do to bring the risk back within risk tolerances. We may need to enhance the existing risk response plan or to enact an altogether different risk response plan.

For example, if we had a risk response plan of *transfer* where we were going to engage a third-party specialist, we need to assess how effective that transfer of risk has been. We need to verify that by transferring the risk to the third-party specialist that the residual/secondary risks have dropped as low as we had anticipated.

If the transfer risk response plan has not reduced the residual/secondary risk to within risk tolerances, we need to determine what we will do to reduce the risk further. Once we have determined what we will do, we need to take action, and then we need to reassess the risk and the risk response plan to ensure that our corrective action has achieved our intended results.

This iterative process follows the PDCA cycle (plan-do-check-act) as per the diagram below.

Diagram 39 – Plan-Do-Check-Act Cycle of Iterative Risk Response Plans

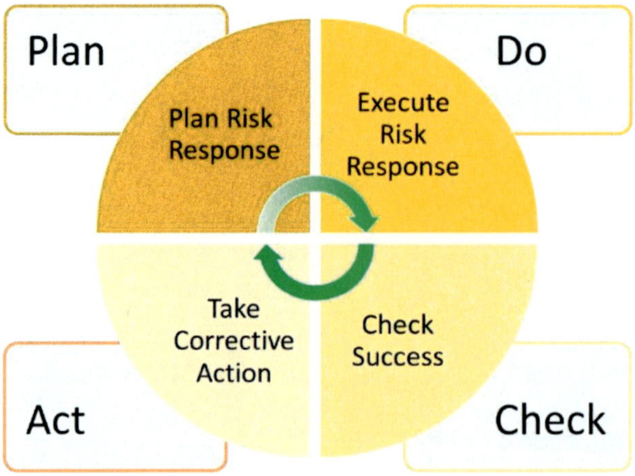

In another example, if we have risk response plans that were intended to mitigate a risk, we need to verify that our risk response plans have actually reduced the impact and/or reduced the probability of the risk occurring. If these risk response plans have not brought the risk levels within risk tolerances, we need to determine what other mitigations or other risk response plans we implement.

9.3.1 Contingency Plans

Contingency plans are risk response plans that will only be executed if a predefined trigger is activated.

Some risk response plans are intended for use only if particular events occur. For some risks, it is fitting for the project team to develop a response plan that they will only implement under specific predefined conditions.[53]

If we have risk response plans that are contingency plans, we want to check on these as well. We want to reassess contingency plans to verify that they are still the most appropriate risk response plans. If we determine that they are not, we may have to modify the contingency plans, replace the contingency plans, or enact another type of risk response plan.

If we have contingency plans whose triggering events have occurred and the contingency plans had been enacted, we need to assess the residual and/or secondary risks to see if they achieved the expected result. Either way, we should update the risk register with the results.

9.4 Iterative Risk Management

Allow me to extend the earlier analogy of an ICT project to the human body. I think most of us would agree that if we went to our doctor for a check-up and we were given a clean bill of health, we would go back to the doctor again the next year for another annual checkup, even if we didn't display any symptoms of illness.

We'd want to keep monitoring our health so that we could address any problems early. We wouldn't wait until something went wrong before going to the doctor for a checkup.

So too with ICT projects. We want to make sure we regularly monitor risks. But "regularly" means something different in project risk context than it does in human medical context. As humans, we have a life that spans many decades, so having a check-up once a year seems a reasonable timeframe.

But, ICT projects have much shorter lives. An ICT project is much more dynamic. It develops and changes much more rapidly than the human body. Because of a project's shorter lifespan and quicker development, we want to do a "risk checkup" much more often. We want to make sure we frequently reassess the project for risks that have changed and for new risks that have emerged.

[53] PMBOK section 11.5.2.6 page 445

How often you reassess risks is dependent on the nature of your project and of your project sponsor's risk appetite. You will have to determine how often you run risk reassessments and how thoroughly you run them.

Some of the project factors to consider when determining this frequency are the size of your project, the complexity of your project, the importance of your project, and the methodology you are using.

ICT projects often run weekly reviews of the urgent risks to ensure that there haven't been any big changes that they need to address. But, a full risk reassessment takes time and effort, so they are usually done less frequently.

Most projects I've worked with perform full risk reassessments somewhere between monthly and quarterly. That frequency range seems to represent an acceptable balance between effort spent on risk management and on time taken away from project delivery.

9.5 Evaluation of the Effectiveness of the Risk Management Process

> *"If the plan doesn't work, change the plan, not the goal."*
>
> Unknown

One of the purposes of the monitor and control process is to scrutinize the effectiveness of the entire risk management process. If we find that the way we had planned our risk management plan isn't proving to be effective, we may need to change the plan.

Early in the project, we determined how we were going to manage risks for this project, and we documented that in the risk management plan. Evaluating the whole risk process effectiveness is about examining the results of our risk management methods and efforts to see how effective our plans to manage risks have turned out to be.

This evaluation process takes a look at the big picture. What did we intend to achieve with our risk management plan? How had we planned to achieve it? How effective have our planned efforts been? Did we execute our plans as we intended? If so, were our efforts effective? Did we achieve the results we had intended?

Once we have assessed the effectiveness of our risk management, we need to decide what we are going to do about it. Was the result of our risk management effort within tolerances? Do we need to change the activities that we engage in to manage risks? Do we need to change when we do them? Do we need to choose how often we do them? Do we need to change how we do them? These are all questions that we must consider as part of the evaluation.

The answers to these questions will be dependent on how far you are in your project lifecycle and how much your performance results differ from what you had planned. Whatever your circumstances, you will almost certainly want to involve other stakeholders in making these evaluations and decisions.

9.5.1 Technical Performance Analysis

Technical performance analysis involves comparing technical accomplishments to the schedule of technical tasks. It requires the definition of objective and quantifiable measures of technical performance, which you can use to compare actual results against targets.

Such technical performance measures may include technical milestones, processing times, number of delivered defects, etc. Deviation can indicate the potential impact of threats or opportunities.

Wherever possible, technical performance analysis should feed into risk analysis and risk reporting throughout the monitor and control process.

9.5.2 Reserve Analysis

During the course of the project, some individual risks will inevitably be realized and turn into issues. When that happens, they will start to consume your schedule and/or budget reserves.

Reserve analysis is the tool we can use to assess how much of our reserves we have consumed and how the amount consumed relates to our expectations.

At any point in the project, we should be able to see how much of our *contingency reserves* we have consumed against our remaining contingency reserve requirements. This will indicate whether we have sufficient contingency reserves to cover our known uncertainties.

Similarly, we can apply reserve analysis to our *management reserves* to assess the adequacy of our management reserves in insulating our project from the unknown, uncertainties.

Unfortunately, our management reserve analysis is restricted to assessing how much of our management reserves we have consumed against how much we have left. There is no way to estimate the need for management reserve requirements to cover the remaining unknown uncertainties.

9.5.3 Risk Audits

Risk audits are a type of formal appraisal that we can use to evaluate the effectiveness of the risk management process as a whole. The project manager is responsible for making sure that risk audits are performed at a suitable frequency, as you defined in the risk management plan.

You might include risk audits during regular project reviews, or you may choose to hold distinct risk audit sessions. The format for the risk audit and its objectives should be unambiguously explained to the relevant stakeholders before the audit is performed.

9.6 Risk Reporting

ICT projects are complex and complicated. Each project objective that we examine might give us a different view of the risks faced by the project. Projects are kind of like the human body.

When assessing the health of the human body, there are many different areas a doctor would look at to assess our health risk. Doctors don't just look for one single measure that would determine our health. Rather, they take measurements of different areas and systems in the body, to get a view of our overall health and to identify any particular areas of concern.

I recently went to my doctor for an annual check-up. He ran through a series of different checks: blood pressure, standing heart rate, cholesterol, weight, temperature, etc. When he gave me the results, they were in a normal range in most areas, but my blood pressure was a cause for concern. Overall, the results were good, but that was one area that needed further investigation.

When we run through risk assessments for ICT projects, we often find similar results. We might find the overall project risk is acceptable. And, we might find that the risks for quality, scope, resources, benefits, and schedule are acceptable. But, we might find that the risks for costs are an area of concern. This doesn't mean that the project is unhealthy, it just highlights areas of specific concern that we may need to address as part of risk management.

In our risk reporting, we should reflect the difference in risk for each of our project objectives. Risk reporting should give an overall assessment of the project risk. But, we should report on the risk of each project objective and highlight the risk of any project objective that is an area of concern, as described in section 5.6.6.

Risk reporting should include objective risk data, supplemented by subjective assessments of the results of quantitative analysis, trend analysis, and reserve analysis. It should provide a subjective assessment, informed by the quantitative metrics. The objective quantifiable risk results should be augmented by a subjective analysis, drawing on your expert experience.

If your doctor handed you a sheet with all your different medical results against the "normal" range for your age, you might be alarmed by any results that lie outside of normal. But doctors don't do that.

Doctors assess the objective medical data and provide a subjective analysis of the data, based on their expert judgement. They show you the data, they explain which outliers they are concerned with. And they explain why they are concerned with those outliers and why they aren't concerned about other outliers. They use their expertise and experience to provide a subjective assessment of the objective data.

We should do the same thing with risk reporting. We need to look through the results of the qualitative and quantitative analysis and apply our own expert judgement to highlight areas of concern. And, if there are any data points that lie outside of acceptable tolerances, we need to explain in our reporting the reasons our level of concern and the reason why we have that level of concern.

Risk reporting isn't about simply presenting the facts. Risk reporting is also about providing a subjective assessment of the facts. We should apply our expert judgement over the facts to explain the areas that we believe should be of high concern. And, for those risks that we believe are of lesser concern, we should explain why.

If our sponsors and project boards are going to make the best-informed decisions, they deserve to be presented with subjective assessments of the risks, not only the objective risk data.

9.6.1 Trend Analysis

We use *trend analysis* to forecast future performance based on past results. Trend analysis looks ahead into the future of the project for expected problems. It warns management of potential problems if current trends persist.

Trend analysis in risk management may be less complex than in cost and schedule trend analysis, but it is no less important. Trend analysis in risk management allows the project manager to forecast project risks.

Trend analysis in risk management can give managers early indications that a project's risk exposure is reducing, remaining static, or increasing. It can give an early indication of a project's risk exposure trends, so we can make informed decisions and take corrective actions.

Trend analysis' biggest benefit is that it gives us an early warning if the risk exposure is increasing. With trend analysis, we don't have to wait until the risk exposure breaches risk thresholds. Trend analysis enables us to wave a flag of warning and take action much earlier.

Importance of Consistency

In order for trend analysis to be accurate, we need to maintain *consistency*. We need to utilize the same risk assessment methodology throughout the project. We need to use the same tools and techniques to gather data, request input, analyze data, and report. We need to use the same tools and techniques to ensure that we are comparing apples to apples when we compare past data to present data.

In an ideal world, we would also have the same people involved throughout the project to ensure that any subjective assessments of risk inputs or of risk analyses are subject to the same, consistent cognitive biases.

Involving the same people in risk management activities is more likely to result in the same cognitive biases being applied, resulting in more consistency and more accurate identification of trends.

Trend analysis enables trend reporting. Trend reporting enables management to make decisions earlier than they would otherwise. Trend reporting enables management to see where the project is heading and to take proactive action if they don't like the trajectory.

Trend reporting allows management to reallocate effort and resources in the face of a risk profile that is trending towards a threshold. Trend reporting enables management to focus more resources on project risks if the overall project risk profile is trending up. Trend reporting also enables management to focus attention and resources towards particular risks and areas of uncertainty that are trending higher.

If you do perform trend analysis and reporting in your risk management, endeavor to use the same methods throughout the project. Whenever possible, try to keep the same stakeholders involved consistently throughout the project. Consistency is critical to realizing the benefits of trend analysis.

The lack of consistency can mislead management with false trend identification. Changing the methods that you use to identify and monitor risk mid-project can skew the results and render trend analysis useless, or worse. False trend identification can either waste time and treasure in chasing shadows (false negatives) or it can disguise the true size of the project's risk exposure (false positives).

If there are inconsistencies in the methods and stakeholders used in tracking trends, there is a danger that we waste time and money chasing phantoms.

Such a misreading could be caused by changing the methods used to assess risk, or it could be caused by different people bringing new cognitive biases to their assessments. Either of which could result in the risk being measured differently than it was measured earlier, despite the real risk exposure remaining unchanged.

This type of false negative can cause a perception that risk exposure is changing when, in reality, it is not. When such false trends are reported, there is a danger that decisions could be made, and actions taken to address an increase in risk that was really only a false negative reading.

Conversely, involving new methods and stakeholders could result in the project measuring risk exposure the same as before, when, in reality, the project's risk exposure has increased.

Reporting such a false positive is even more dangerous than reporting a false negative. Such reporting can cloud management's view of the project's increasing risk exposure. Management may make decisions with a rosy view of the level of risks involved, unaware of the growing danger lurking just below the waterline. This can lull management into a false sense of security, when, in reality, disaster may be just around the corner.

Conclusion

Trend analysis and trend reporting are tools that can enable projects to respond early to changes in their risk exposure. This early response can save a project time and money and increases the project's chances of success.

But, trend analysis works best when we maintain consistency in the methods and people, we involve in the risk assessment, trend analysis, and trend reporting.

9.7 Chapter Summary – Monitor and Control

In this chapter, we've covered the reasons why we monitor risks, the activities involved in monitoring risks, and the tools and techniques to use in monitoring risks. In this chapter we've covered

- Identify and analyze emergent risks
- Track identified individual risks
- Monitor the implementation of risk responses
    - Contingency plans
- Iterative risk management
- Evaluation of the effectiveness of the risk management process including
    - Technical performance analysis

    - Reserve analysis
    - Risk audits
- Risk reporting including risk trend analysis and trend reporting

Final Words

Congratulations on finishing this book! I hope that you feel confident and well prepared to manage risks in your ICT project.

I hope this book has been of value to you and I hope it helps contribute to your success. Good luck with your project and with your career!

If you are interested in an editable and printable pack of the tools and material in this book, please go to www.ictrisk.com. I've put together a pack with all the tools and templates you need to manage risks in an ICT project. The pack should make your job easier and it will make your efforts look professional. Why dedicate the time and effort into reinventing the wheel when I've created a perfectly good wheel for you!?!

If you would like to make suggestions or comment on any of the material in this book, please go to www.ictrisk.com. Thank you; I appreciate your feedback. I look forward to improving the next edition of this book with your contribution.

If you leave a positive review on Amazon, please leave your name in the review. I will be sure to include your name in the acknowledgement section in the next edition of this book to show my appreciation!

Many Thanks!

Alan Bergsma

A Request

I sincerely hope that you have found this book helpful. If you have, I would be grateful if you would leave a positive review on Amazon. Please include the title, *Bullet Dodging - A Risk Management Handbook for ICT Projects: A Risk Management Handbook for ICT Projects*, and please share what you found useful in the book. And please leave your name so that I can thank you!

If you have any criticism, I'd love to hear about it, so I can improve the next edition of this book. Please forward any criticism via www.ictrisk.com so that I can add your contribution to the next edition of this book.

Many thanks!

Alan Bergsma

Appendix A: Risk Categories for the Workshop

Executive Support	Resources & Teams	Decision Making	Project Management	Environmental	Contracts
Scope	Architecture	Procurement	Secondary Risks	Estimates	Suppliers & Vendors
Cost Management	Design	Authority	User Acceptance	Interfaces	Client / Customer Stability
Organizational Change Management	Technical & Technology	Approvals & Red Tape	Commercial	Program / Portfolio Management	Partnerships & Joint Ventures
Stakeholders	Integration	Organizational	Security	Operations	Legislation & Regulation
Communication	Requirements	External Risks	Legal	Onboarding	Tariffs & Export Controls
Exchange Rates	Facilities	Change Freezes	Leadership & Vision	Change Control	Environments & Release Management

Appendix B: Checklist – Probing Questions by Category [54]

Use the examples below to flesh out the categories when you probe your participants to ensure that all avenues have been considered.

Executive Support

1. **Executive support fails**
   The project team may lack the authority to achieve project objectives without the sponsor. In such cases, executive management support is essential to project success. When this disappears, the project fails.

2. **Executives become disengaged with project**
   Executive management disregards project communications and the project schedule and costs blow out while waiting for approvals.

3. **Conflict between executive stakeholders disrupts project**
   Members of executive are combative with the project team or there is a disagreement over project issues at the executive level which disrupts progress.

4. **Executive turnover disrupts project**
   A key executive leaves the company or changes roles disrupting progress.

Scope

5. **Scope is poorly defined**
   There is a general risk of an error or omission in scope definition.

6. **Scope-creep inflates scope**
   Uncontrolled changes and continuous growth of scope cause budget and schedule blowouts.

7. **Gold-plating inflates scope**
   The project team adds additional product or service features that are not in the requirements or change requests.

8. **Estimates are inaccurate**
   Inaccurate estimates are often a project risk.

9. **Dependencies are inaccurate**
   Dependencies dramatically impact the project schedule and costs and can be difficult to assess accurately. Are there any missing dependencies? Are the impacts of identified dependencies fully understood?

[54] https://management.simplicable.com/management/new/130-project-risks

10. **Activities are missing from scope**
    Required activities are missing from scope definition.

Cost Management

11. **Cost forecasts are inaccurate**
    Inaccurate cost estimates and forecasts are often a source of risk.

12. **Exchange-rate variability**
    When costs are incurred in foreign currencies, exchange rates can have a dramatic impact.

Organizational Change Management

13. **Lack of change-management system**
    Identify any lack of critical tools, capability, or capacity as a risk.

14. **Lack of change management process and communication**
    Change management is critical to project success.

15. **Change fatigue within the organization**
    An organization's ability to accept change is inversely related to the amount of change they are required to accept.

Stakeholders

16. **Stakeholders become disengaged**
    Stakeholders ignore project requests or communications.

17. **Stakeholders have inaccurate expectations**
    Stakeholders develop erroneous expectations (believe that the project will achieve something not in the requirements, plan, etc.).

18. **Stakeholder turnover**
    Stakeholder turnover can cause misunderstandings and delays that lead to project disruptions.

19. **Stakeholders withhold support for project**
    When stakeholders have a negative attitude towards the project and/or wish to see it fail.

20. **Stakeholder conflict**
    Disagreement between stakeholders over project issues cause delay.

21. **Process input quality is low**
    Low-quality inputs from stakeholders.

Communication

"The single biggest problem in communication is the illusion that it has taken place."

George Bernard Shaw, Irish playwright, critic, polemicist, and political activist

22. Project team misunderstand requirements
When requirements are misinterpreted by the project team, a gap develops between expectations and delivery.

23. Communication overhead
When key project resources spend a high percentage of their time engaging stakeholders on project issues and change requests, their work may fall behind the schedule.

24. Lack of communication
Until messages are confirmed as received and clearly understood, there may be miscommunications.

25. Users have inaccurate expectations
There is a risk that inaccurate expectations cause user resistance and lack of cooperation.

If users believe the project is building an apple when you're really building an orange (i.e., users don't understand the product or service that's coming their way).

26. Impacted individuals are not kept informed in a timely manner
A stakeholder is missing in your communication plan. Anyone who is not informed but is impacted may create roadblocks.

Resources and Team

27. Resource shortfalls
Not able to secure sufficient resources for the project.

28. Learning curves lead to delays and cost overruns
When the project team needs to acquire new skills for the project, there is a risk that productivity will be low or slow.

29. Training is not available
Quality training for certain skills can be difficult to secure.

30. Training is inadequate
Training is often a poor substitute for professional experience. Projects should not assume that resources will be fully productive in a new skill.

31. Resources are inexperienced
Resources that are new to your industry or profession tend to make more mistakes and be less productive.

32. Resource performance issues
Resource people perform below expectations.

33. Team members with negative attitudes towards the project
People who are negative towards the project may actively or passively sabotage project efforts.

34. Resource turnover
Resource turnover leads to delays and cost overruns.

35. Low team motivation
Your team lacks motivation. This is a particularly common risk for long-running projects.

36. **Lack of commitment from functional managers**
    In a matrix organization, your team may report to functional managers. These functional managers are important stakeholders whose support is critical.

Architecture

37. **Architecture fails to pass governance processes**
    Plan for any architectural or technology governance processes that the project may need to pass.

38. **Architecture lacks flexibility**
    The architecture is incapable of supporting change requests and needs to be reworked.

39. **Architecture is not fit for purpose**
    The architecture quality is low.

40. **Architecture is infeasible**
    The architecture is impossible to implement, excessively costly, or does not support the requirements.

Design

41. **Design is infeasible**
    The design is flawed, is excessively costly, or does not support the requirements.

42. **Design lacks flexibility**
    Poor design makes change requests difficult and costly.

43. **Design is not fit for purpose**
    The design quality is low.

44. **Design fails peer review**
    It is good practice to have peers or architectural experts review the designs.

Technical & Technology

45. **Technology components are not fit for purpose**
    Technology component quality is low.

46. **Technology components are not scalable**
    Components can't be scaled to meet performance demands.

47. **Technology components are not interoperable**
    Components lack standard interfaces.

48. **Technology components are not compliant with standards and best practices**
    Non-standard components violate best practices.

49. **Technology components have security vulnerabilities**
    Security vulnerabilities are key technology risks.

50. **Technology components are over-engineered**
    A component may be bloated with unneeded functionality and design features.

51. Technology components lack stability
Equipment fails.

52. Technology components are not extensible
It may be difficult to add capabilities to existing components.

53. Technology components are not reliable
Components may fail after a short time.

54. Information security incidents
There is the risk of a security incident during the project.

55. System outages
Critical systems, such as your test environments, go down.

56. Legacy components lack documentation
Integration with undocumented legacy components is a high-risk activity.

57. Legacy components are out of support
There is a risk of integrating with legacy components that are no longer supported.

58. Components or product or services are not maintainable
Technology components, tools, or platforms are difficult to maintain.

59. Components or product or services cannot be operationalized
Technology operations may have criteria for operationalization of new systems that need to be met.

60. Project management tool problems
There are technical problems with the project management tools themselves.

Integration

61. Delays to required infrastructure
There is a risk of delays to infrastructure such as hardware or software.

62. Failure to integrate with business processes
There is a risk that your product or service fails to fit into the existing business.

63. Failure to integrate with systems
There is a risk that your product or service will fail to integrate with existing systems.

64. Integration testing environments are not available
There is a risk that environments will not be available to test integration.

65. Failure to integrate with the organization
There is a risk that your project does not integrate with the organization. This happens when the project is focused on delivering something specific and fails to look at the organization as a whole.

66. Failure to integrate components
There is a risk that product or service components will fail to integrate with each other. This can represent a significant risk when you've outsourced work to a large number of vendors.

67. Project disrupts operations
There is a rick that the project will disrupt business operations and damage the firm's financial results.

68. **Project disrupts sales**
    There is a risk that the project disrupts sales effectiveness.

69. **Project disrupts compliance**
    There is a risk that the project disrupts compliance processes, such as audits and reporting.

Requirements

70. **Requirements fail to align with strategy**
    Your requirements conflict with the firm's strategy. If you sense that this is the case, list it as a risk.

71. **Requirements fail to align with business processes**
    The requirements make no sense in the context of the business.

72. **Requirements fail to align with systems**
    The requirements do not align with other systems.

73. **Requirements have compliance issues**
    If you have any doubt that requirements comply with the law, list it as a risk.

74. **Requirements are ambiguous**
    Requirements are unclear and open to interpretation.

75. **Requirement quality is low**
    Requirements are not fit for the purpose.

76. **Requirements are incomplete**
    You can spot obvious holes in the requirements.

Decision Making and Issue Resolution

77. **Decision delays impact project**
    Establish guidelines for decision turnaround time. Identify the risk that guidelines will be exceeded.

78. **Decisions are ambiguous**
    Stakeholders may tend to make decisions that are intentionally ambiguous (a responsibility avoidance technique). This can be identified as a risk and managed.

79. **Decision quality is low**
    Decisions are not fit for the purpose.

80. **Decisions are incomplete**
    There is a risk that issue resolutions either do not address the issue or create more issues.

Procurement

81. **No response to RFP**
    There is a risk that there is limited response to an RFP. This occurs when the RFP terms are unacceptable to vendors or if your firm has a bad reputation among vendors.

82. **Low-quality responses to RFP**
    There is a risk that you receive half-hearted (i.e., unusable) responses to your RFP.

83. Failure to negotiate a reasonable price for contracts
There is a risk that you are not able to negotiate a reasonable price for contracts. This occurs when the requirements or contract terms make vendors nervous.

84. Unacceptable contract terms
There is a risk that you are unable to negotiate acceptable contract terms.

85. Conflict with vendor on project issues
The relationship with a vendor turns to conflict and project issues mount.

86. Conflict between vendors on project issues
Your vendors develop conflicts with each other, and cooperation breaks down.

87. Vendor starts late
There is a risk that the vendor's schedule delays the project start.

88. Vendor components fail to meet requirements
A vendor misunderstands requirements or delivers components that do not meet the requirements.

89. Quality of vendor components is low
Vendor outputs are not fit for the purpose.

90. Infrastructure quality is low
The infrastructure fails or is not fit for the purpose.

91. Service quality is low
Services you procure, such as consulting, are not fit for the purpose.

92. Vendor components introduce third-party liability
Vendor components introduce liability (e.g., they violate patents).

93. Loss of intellectual property
Vendors use your IP for their own purposes.

Authority

94. Project team lacks authority to complete work
If you lack specific authorities required to deliver the project, list this as a risk.

95. Authority is unclear
It is unclear who has the authority to accomplish a project objective.

Approvals and Red Tape

> *"Most of what we call management consists of making it difficult for people to get their work done."*
>
> Peter Drucker - American management consultant, educator, and author

96. Delays to stakeholder approvals impact the project
There is a risk that approval deadlines are exceeded.

97. Delays to financial approvals impact the project
There is a risk of delays to financial approvals and processes to release funds.

98. Delays to procurement processes impact the project
Many organizations have specific procurement processes that must be followed. These processes can be time-consuming and highly variable. You may want to list the procurement process as a risk.

99. Delays to recruiting processes impact the project
If your project involves recruiting resources, this may take many months and is highly variable.

100. Delays to training impact the project
If your training budget requires separate approvals (e.g., from functional managers or HR), document the risk that this will be slow.

Organizational

"Organization charts and fancy titles count for next to nothing."

Colin Powell, American politician, diplomat and retired four-star general

101. The project fails to match the organization's culture
There may be a culture fit issue between your product or service and the organization. For example, if the organization's culture calls for employees to bring their own mobile devices to work (BYOD) and you build a user interface that only works on corporate devices.

102. An organizational restructuring throws the project into chaos
If your project has a large footprint, it may be extremely sensitive to other organizational changes.

103. A merger or acquisition disrupts the project
Mergers and acquisitions may represent significant organizational changes.

External

104. Legal and regulatory change impacts project
If your project spans areas that are compliance sensitive, you may want to list regulatory change as a risk.

105. Force majeure (e.g., act of nature) impacts project
There is a risk that major disruptions, such as a natural disaster, impact your project.

106. Market forces impact project
Market changes may impact a project (e.g., a market crash).

107. Technical change impacts project
A technology innovation changes your industry and impacts the project.

108. Business change impacts project
A business innovation changes your industry and impacts the project.

Project Management

109. Failure to follow methodology
If your organization asks you to change your project management methodology, the disruption would be a risk.

110. Lack of management or control
A lack of project management should be documented as a risk. For example, if resource constraints cause the project to skip certain project management best practices.

111. Errors in key project management processes
There is a risk of errors in project management, e.g., schedule errors impact your project.

Secondary Risks

112. Counterparty risk
The risk of a third party not delivering / acting as agreed. This could be a vendor, supplier, insurer, financial institution, manufacturer, joint venture partner, or any other third party.

User Acceptance

113. Users reject the prototype
One of the key methods of improving user acceptance is to get regular prototypes in front of users. There's always a risk that these prototypes will be rejected (i.e., will require significant rework).

114. User interface does not allow users to complete tasks
There may be a risk that the user interface does not allow users to complete end-to-end tasks.

115. User interface is low quality
The user interface is slow or difficult to use.

116. User interface is not accessible
In many jurisdictions, user interfaces *must* be accessible (e.g., employment or consumer law). Many organizational cultures require accessible user interfaces.

117. Project reduces business productivity
Users identify your product or service as reducing their productivity.

118. Project reduces innovation
Users identify your product or service as a roadblock to innovation.

119. Product or service disrupts business metrics (measurements of objectives)
Your product or service launches causes business KPIs to worsen. For example, if you launch a new ERP and the supply-chain cycle times jump.

120. Users reject the product or service
There is a general risk that users will reject your product or service.

Commercial

The following risks may apply to new ICT product or service development projects.

121. Product or service does not sell
There may be a demand risk for the new product or service.

122. Product or service incurs legal liability
The product or service has quality issues that harm your customers.

123. Product or service negatively affects brand
The product or service has quality issues that damage your brand.

124. Product or service negatively affects reputation
The product or service generates negative publicity and/or damages customer relationships.

Security

125. Project negatively impacts security
The project introduces new security threats.

126. Project negatively impacts security controls
The project reduces the existing protection and exposes the organization to threats.

Legal

127. Project negatively affects management
The project exposes management to legal action due to a lack of compliance.

128. Project exposes the organization to liability
The actions taken by the project cause an external party to sue for damages.

Environmental

129. Project causes a regulator to revoke the license to operate
The project breaches environmental regulations and causes the regulator to revoke the license to operate.

130. Project negatively affects reputation
The project causes an environmental impact that generates negative publicity.

Estimates

131. Actual costs are higher than estimates
The costs turned out to be higher than initially anticipated.

132. Actual costs are lower than estimates
The costs turned out to be lower than initially anticipated.

133. Actual amount of time required is higher than estimates
The time required turned out to be higher than initially anticipated.

134. Actual costs are lower than estimates
The time required turned out to be lower than initially anticipated.

Interfaces

135. Data differences make integration difficult
Data fields often do not map well between systems. Even fields that have the same name can have a different definition. And, the field definition can be different from the way the business uses the data field. Data mapping requires interpretation and often takes longer and costs more than anticipated.

136. Integration confirmation and error protocols are not agreed
Assumptions on integration protocols can lead to errors unless they are specifically defined and agreed between the interfacing parties.

137. Lack of defined change control for interfaces
Uncontrolled changes made to one side of the interface can cause the interface to fail if the changes are not mirrored on the other side of the interface.

Program/Portfolio Management

138. Risk escalation process to escalate risks to program/portfolio are not defined
Undefined escalation processes can cause confusion and delays.

139. Risk delegation process to delegate risks from program/portfolio are not defined
Undefined delegation processes can cause confusion and delays.

140. Issue escalation process to escalate risks to program/portfolio are not defined
Undefined escalation processes can cause confusion and delays.

Operations

141. Changes to the nature of business operations could change the nature of the business requirements
Any changes to the nature of the work that operations engage in could cause changes to the business requirements.

142. An increase in business demand reduces SME availability
Any increase in the demands of operational activities (such as a spike in operations) could limit the project's ability to access SMEs for advice and decisions.

Onboarding

143. A protracted onboarding process for new recruits causes delays to the schedule
Often there are procurement and human resources procedures that a project has to go through before bringing new people into the team. This can cause delays to the schedule and incur additional costs.

144. Onerous onboarding processes can deter the best talent
The people who are very good at what they do often have a choice of many different jobs. If the onboarding process in your organization is too difficult for the person coming in, the most talented people will simply go elsewhere.

Contracts

145. The contracts are ambiguous and introduce uncertainty
In projects, contracts are often written before we understand all the details. As we discover more details, ambiguous contracts introduce uncertainty and can result in issues being decided in court, causing project delays.

146. The final contracts have been negotiated by the procurement department who do not understand the implications of ICT components in the project
When procurement negotiates ICT projects, they do not always understand the implications of the clauses that they negotiate in/out. This can result in a change to scope, increase in cost, reduction of quality, or a delay in schedule.

Suppliers/Vendors

147. Suppliers / vendors introduce counterparty risk
Suppliers and vendors may be outside of a project's direct control. They are subject to unknown influences and may or may not act according to expectations.

148. Suppliers / vendors commercial viability may be in question
As a third party, suppliers and vendors are not obligated to disclose their commercial viability. If they are under financial stress, the quality and speed of delivery may be lower than expected. And, if the financial stress is bad enough, they may go bankrupt.

Client/Customer Stability

149. Your client / customers may undergo a change that disrupts your project
If your client / customers get caught up in a restructuring or other disruptive organizational activity, your project may not be able to access SMEs or decision makers and may be deprioritized or cancelled.

150. Your client / customers undergo a significant change that produces change fatigue in your project's future users
When organizations endure significant organizational changes, users can become fatigued enough that they resist further change. This can impact your project's success in implementing your organizational change activities and in implementing your solution.

Partnership and Joint Ventures

151. Partnership and joint ventures introduce counterparty risk
Partnership and joint ventures are outside of a project's direct control. They are subject to unknown influences and may or may not act according to expectations.

152. Partnership and joint ventures commercial viability may be in question
As a third party, partnership and joint ventures are not obligated to disclose their commercial viability. If they are under financial stress, the quality and speed of delivery may be lower than expected. And, if the financial stress is bad enough, they may go bankrupt.

Legislation and Regulation

153. New legislation or regulation could be introduced that impacts your project
Legislation or regulation changes could impact any aspect of your project effecting scope, quality, cost, or schedule.

154. Changes to the enforcement of legislation or regulation could impact your project
Legislation and regulation can be enforced to differing degrees of effectiveness. If a government increases or decreases the rigor to which they enforce legislation and regulation, it could impact your project scope, quality, cost, or schedule.

Exchange Rates

155. Changes to exchange rates could impact your costs if you're buying goods or services outside of your local currency
A swing in exchange rates could change the costs of any goods or services that you're buying outside of your local currency.

156. Changes to exchange rates could impact the quality of services delivered by overseas suppliers
A swing in exchange rates could reduce the profitability of your supplier's contract with your project. If your project becomes unprofitable for your supplier, the quality and speed of their services may decline.

Tariff and Export Controls

157. Tariffs can introduce additional costs for procuring international goods or services
A swing in exchange rates could change the costs of any goods or services that you are buying outside of your local currency.

158. Export controls can disrupt the supply of goods and services
Export controls may mean that you cannot source goods and services from the country you had intended. This may entail sourcing goods and services from elsewhere, which can introduce delays to schedules, increased costs, and reduced quality.

Facilities

159. Project facilities are so unpleasant that they become a hygiene factor that makes it difficult to attract and retain talented resources
Projects are often squeezed into whatever facilities are available, regardless of how unpleasant. Unpleasant working environments can make it difficult to attract and retain talented resources.

160. Project teams are not co-located due to facility restrictions, reducing team productivity
Physical distance impedes communication and collaboration. A team's effectiveness is reduced by even the smallest of distance separation.

Change Freezes

161. Organizations can enact change freezes that cause delays to projects
Organizations may enact change freezes for any number of reasons. Change freezes that occur during release management activities can cause delays and disrupt your schedule.

162. Change freezes can lead to cascading delays
A change freeze may only delay a project by a couple of weeks, but if there are additional schedule dependencies, the resulting delay could grow exponentially.

Leadership and Vision

"All things are created twice; first mentally; then physically. The key to creativity is to begin with the end in mind, with a vision and a blueprint of the desired result."

Stephen Covey, American educator, author, businessman, and keynote speaker

163. A lack of leadership can slow project progress and dull team moral
Projects need leaders to guide team progress. The leaders do not need to hold any title; they just need to help inspire the team to achieve the project objectives. A lack of leadership dulls team morale and saps productivity.

164. A lack of vision results in scope creep, gold plating, and rework
A lack of vision deprives the project team of guidance on what is in scope and out of scope. It can result in effort expended on non-critical work, and it can take effort away from critical work, resulting in lower quality, inflated scope, increased costs, and schedule delays.

Change Control

"When it is obvious that the goals cannot be reached, don't adjust the goals, adjust the action steps."

Confucius

165. Change request overload
A large number of change requests **dramatically** raises the complexity of the project and distracts key resources reducing team efficacy.

166. Stakeholder conflict over proposed changes
Change requests may be the source of stakeholder **conflict**.

167. Perceptions that a project failed because of changes
Large numbers of high-priority change requests may lead to the **perception** that the project has failed. When the schedule and budget are continually extended, stakeholders may feel the project missed its original targets.

168. Change request quality is low
Change requests are low-quality (e.g., **ambiguous**).

169. Lack of change control process
A change control board or similar mechanism is **essential** to managing change for large projects.

170. Inaccurate change priorities
Non-essential changes are prioritized, impacting **critical** schedules.

171. Change request conflicts with requirements
Change requests clash with the **requirements**.

172. **Lack of a change control mechanism can result in scope changes that result in promised benefits not being delivered**
Lack of change control can result in the project dropping **features** out of scope that were required to achieve the business case benefits.

173. **Lack of a change control mechanism can result in inflated scope delivery beyond what was agreed**
Lack of change control can result in the project adding features beyond what is required to achieve the business case benefits.

174. **Lack of adherence to a change control mechanism can result in scope changes that result in promised benefits not being delivered**
Lack of adherence to change control mechanisms can result in the project dropping features out of scope that were required to achieve the business case benefits.

175. **Lack of adherence to a change control mechanism can result in inflated scope delivery beyond what was agreed**
Lack of adherence to change control mechanisms can result in the project adding features beyond what is required to achieve the business case benefits.

Environments and Release Management

176. **Lack of an effective release management mechanism can result in confusion and rework**
When the release management mechanism is non-existent or is ineffective, delays, confusion, and rework are likely.

177. **Lack of adherence to a release management mechanism can result in confusion and rework**
When the release management mechanism is not adhered to by the project or other stakeholders, delays, confusion, and rework are likely.

Appendix C: Reticular Activating System's Impact on Project Management

Our eyes see millions of things every day. Our ears hear hundreds of thousands of words and sounds every day. Our *reticular activating system* (RAS) protects our brains from overload. Even with our RAS in action, some days it can feel like we are absorbing an overwhelming amount of information. Can you imagine what life would be like if we didn't have a RAS to limit the amount of information we take in? Complete information overload!

Even if you didn't know what it was called, we've all seen the RAS in action. Have you ever considered buying a particular model of car and then noticed that suddenly you see that model everywhere?

It's not that the model has suddenly become more popular. That model has always been there, all around you. But once you started to focus on it, your RAS sets up a filter for it. And then your RAS presents your brain with the data every time your eyes see it.

You've probably seen the RAS in action with other people's dispositions as well. Some people focus on the positive and always seem to see the bright side of any situation. Their proverbial glass is always half full. Conversely, some people always see the negative in even the most positive situations. Their proverbial glass is always half empty. They both see what they focus on. This is another example of the RAS at work.

Because of the RAS, what we focus on tends to be where our actions take us. If we are constantly focusing on the negative, we can subconsciously engage in actions that lead to negative results.

Project managers face the same danger. We want to make sure PMs are aware of risks, but we want to prevent them from becoming too immersed in risks, as there is a danger that they will subconsciously steer towards the negatives that they are trying to avoid.

As a PM, you want to keep your subconscious positive in order to stay focused on successful project delivery. You don't want your subconscious to fixate it's focus on what could go wrong. You don't want your subconscious encouraging you to engage in actions that could lead to the negative results you are consciously trying to avoid!

As project professionals, we can limit the amount of time we focus on risk by outsourcing much of the risk management function to specialist 3rd party providers. When we limit the time that we spend focusing on negative risk, we can nurture the positive mindset that increases our likelihood of achieving success. By training our focus on the positive, we can harness the power of the RAS and drive towards delivering our projects successfully.

Acknowledgements

First, I would like to thank Frank Teunissen, Adam Chow, Andrew McFarlane, John Lynch, Steve Torokfalvy and Ellen Liashenko for their review and feedback on early drafts of the book. Thank you for your time and expertise. I am grateful.

Second, I would like to thank Luanne Thibault for her proofreading and her help with typesetting / formatting. Quality work as always; thank you.

Third, I would like to thank Anne Reilly for her efforts in copy editing. I've never seen so much red ink! I thank you for your help.

Fourth, I would like to thank some of the professionals I've worked with over the years. They have helped me develop a richer understanding of how to manage risk in ICT projects. I'd like to thank: Tom Fordyce, David Winters, John Meehan, Scott Boyle, Simon Cleary, Helen Moore, Clinton Parker, Jason Hitchcock, Scott Clements, Brendon Taylor, Shaun McKernie, Jeff Jones, Sharon Herridge, Roddy Beck, Ian Hatton, Nicki Kons, John Bowden, Graham Saunders, Michael Saleh, Steve Hands, Mark Watt, Mark Coleman, Maria Zavros, Peter Fox, Samuel Godfrey, Jacqui Elliott, Shane Marshall, Michelle Spencer, Greg Watts, Cristina Lopez, Mervin Chiang, Stephanie Ning, Natasha Hinrichsen, Drew Ellem, Aman Sandhu, Peter Day, Karolyn Gainfort, Daniel Benns, Chris Iddon, Melissa Occhipinti, Sarah Yip, Destiny Flaherty, Michael Meers, Conrad Hall, Andrew O'Connor, Mick Brown, Sabina Gorycka-Pajak, Christine Crain, Sonal Khare, Sarah Sawyer, Rob McGuire, Lee Grindley, Anne Dunstone, Dean Robertson, Dan Cooke, Michael Baker, David Smyth, Andrew Casper, Nadine Harden, Stephanie Houle, Thomas Behe, Shannon Ryan, Mark Elms, Scott Baker, Dan Barr, Ray Moller, Eoin O'Dwyer, Mike Irving, Gerard Creedon, John Robertson, Ben Pace, Craig Favilla, Samara Teefy, Suraj Sood, Riyaz Syed, Warwick Peel, Ross Francis, Sandeep Kapur, Mike Rothery, Ian Davies, Ehsan Alvi, Maxime Groenwoud, Jody Meier, Warren Ryder, Austen Stow, Roddy Beck, George Grubetic, Timothy Beel, Sarie Preiss, Paul McIntosh, Marianne Lucy-Head, Troy Middlebrook, Beverly Dummermuth, Trina Bristowe, Steffenie Langby, Leanne Swan, Russel Withers, Joy Quinn, Stephanie Blouin, Nancy Reeve, Kathryn von Frankenberg, Michelle Nicolay, Matt McDonald, Andrew McMutrie, Jenny Potts, Heath Roberts, Andrew Bruce, James Hallam, Nicholas D'arcy-Fox, Bernie Woodcroft, Scott Ives, Geta Hutchinson, Paul Betts, Christopher Balmain, Sally Watts, Dave Cashmore, Jacob Widerberg, Robin Jones, Leanne Ford, Richard Lloyd, Judith Evans, Andy Leslie, Sean Leonard, Katy Pocklington, Andrew O'Driscoll, Michelle Yorath, Iain Lockhart, Jeff Cruikshank, Kesh Sharma, Mike Lawson, Philip Brown, David Morris, Charles Edwards, Brian Cavanagh, Bruce Stewart, Mike Dearden, Lisa Jackson, Todd Dempsey, Catherine Hargreaves, Peter Wilde, Simon McFarlane, Michael Smith, Marianne Garden, Tina Adams, Darren Blaney, Peter Brown, Nathan Burgess, Mat Byrne, Greg Clarke, Leon Coate, Atul Deshpande, Campbell Dickson, Nicole Doogan, James Fenton, Scott Garner, Matthew Giles, James Hurburgh, Pete Ingram, Bill Keating, Chris Keegan, Marni Manning, Michael Parsons, Daniel Ryan, Gillian Ryder, John Waugh, Annette Wilson, Emma Windle, Jacob Broman-Minish, Vesna Andric, Jacinta Trudinger, Garry Kerr, Mick O'Dowd, and Natasha Lamont.

Bibliography

Axelos. (2010). *Management of Risk: Guidance for Practitioners - 3rd Edition.* Belfast: TSO.

Axelos. (2015). *Prince2 Agile.* TSO.

Axelos. (n.d.). *Managing Successful Programmes .* TSO - Fourth Edition (2011).

Branson, R. (2014, February 25). *Richard Branson - Founder at Virgin Group.* Retrieved from LinkedIn: https://www.linkedin.com/pulse/20140224234637-204068115-best-advice-protect-the-downside/

Bright Hub PM. (2009, January 19). *Stakeholder Analysis - Spheres of Influence - Power, Attitude, Interest.* Retrieved from Bright Hub PM: https://www.brighthubpm.com/project-planning/23481-stakeholder-analysis-spheres-of-influence/

Cialdini, D. R. (2009). *Influence: The Psychology of Persuasion.* USA: HarperCollins - US; Revised ed. Edition .

Deloitte. (2017, September 7). *Project Risk Management Applying the Three Lines of Defence Model to Project Risk Management.* Retrieved from Deloitte: https://www2.deloitte.com/content/dam/Deloitte/au/Documents/risk/deloitte-au-rs-project-risk-management-310117.pdf

Green Book Supplementary Guidance: Risk. (2006, November). Retrieved from Thinking about risk - Managing your risk appetite: a practitioner's guide: https://assets.publishing.service.gov.uk/government/uploads/system/uploads/attachment_data/file/191520/Managing_your_risk_appetite_a_practitioners_guide.pdf

HM Treasury. (2004). *The Orange Book - Management of Risk - Principles and Concepts.* HM Treasury.

HM Treasury. (2006, November). *https://www.gov.uk/government/publications/green-book-supplementary-guidance-risk.* Retrieved from Thinking about your risk - Setting and Communicating your Risk Appetite: https://assets.publishing.service.gov.uk/government/uploads/system/uploads/attachment_data/file/191519/Setting_and_communicating_your_risk_appetite.pdf

HM Treasury. (2009, July). *https://www.gov.uk/government/publications/green-book-supplementary-guidance-risk.* Retrieved from Risk Management assessment: https://assets.publishing.service.gov.uk/government/uploads/system/uploads/attachment_data/file/191516/Risk_management_assessment_framework.pdf

HM Treasury. (2015, March). *The aqua book: guidance on producing quality analysis for government.* Retrieved from assets.publishing.service.gov.uk: https://assets.publishing.service.gov.uk/government/uploads/system/uploads/attachment_data/file/416478/aqua_book_final_web.pdf

HM Treasury. (2016, March). *https://www.gov.uk/government/publications/audit-committee-handbook.* Retrieved from Audit and risk assurance committee handbook: https://assets.publishing.service.gov.uk/government/uploads/system/uploads/attachment_data/file/512760/PU1934_Audit_committee_handbook.pdf

HM Treasury. (2018). *SUPPLEMENTARY GREEN BOOK GUIDANCE .* Retrieved from https://www.gov.uk/government/publications: https://assets.publishing.service.gov.uk/government/uploads/system/uploads/attachment_data/file/191507/Optimism_bias.pdf

HM Treasury. (2018). *THE GREEN BOOK - Central Guidance on Appraisal and Evaluation.* London: HM Treasury.

HM Treasury, Office of Government Commerce. (2004). *https://www.gov.uk/government/publications/green-book-supplementary-guidance-risk.* Retrieved from Managing risks with delivery partners: https://assets.publishing.service.gov.uk/government/uploads/system/uploads/attachment_data/file/191515/Managing_risks_with_delivery_partners.pdf

International Standards Organisation. (2018). ISO 31000:2018 Risk management — Guidelines. *ISO*, 16.

Mar, A. (2016, June 28). *130 Project Risks (List)*. Retrieved from Simplicable: https://management.simplicable.com/management/new/130-project-risks

Michael Bloch, Sven Blumberg, and Jürgen Laartz at McKinsey. (2012, October 1). *Delivering large-scale IT projects on time, on budget, and on value*. Retrieved from McKinsey Digital Insights: https://www.mckinsey.com/business-functions/mckinsey-digital/our-insights/delivering-large-scale-it-projects-on-time-on-budget-and-on-value

Oakes, G. (2008). *Project Reviews, Assurance and Governance.* Burlington, VT: Ashgate Publishing Company.

Praxiom. (2018, August 7). *ISO 31000 2018 Plain English Terms*. Retrieved from Praxiom.com: http://praxiom.com/iso-31000-terms.htm

Project Management Institute. (2009). *Practice Standard for Project Risk Management.* Project Management Institute.

Project Management Institute. (2017). *A Guide to the Project Management Body of Knowledge (PMBOK Guide) - English, 6th Edition Paperback.* Project Management Institute; Sixth Edition (30 September 2017).

Project Management Institute. (2017). *Agile Practice Guide.* Project Management Institute.

Project Management Institute. (2019). *The Standard for Risk Management in Portfolios, Programs and Projects.* Project Management Institute.

Riskamp. (2020). *What is Monte Carlo Simulation?* Retrieved from Riskamp: https://www.riskamp.com/files/RiskAMP%20-%20Monte%20Carlo%20Simulation.pdf

Smartdraw. (2020). *Influence Diagram*. Retrieved from Smartdraw: https://www.smartdraw.com/influence-diagram/

The Standish Group International. (2015). *The Chaos Report 2015.* Retrieved from CHAOS Reports: https://www.standishgroup.com/sample_research_files/CHAOSReport2015-Final.pdf

Thwink.org. (2014). *Causal Chain.* Retrieved from Thwink.org: https://www.thwink.org/sustain/glossary/CausalChain.htm

Wikipedia. (2020, 09 07). *Cognitive shifting*. Retrieved from Wikipedia: https://en.wikipedia.org/wiki/Cognitive_shifting

Wikipedia. (2020, August 27). *Five Whys*. Retrieved from Wikipedia: https://en.wikipedia.org/wiki/Five_whys

Wikipedia. (2020, August 31). *Groupthink*. Retrieved from Wikipedia: https://en.wikipedia.org/wiki/Groupthink

Wikipedia. (2020, August 16). *Optimism Bias*. Retrieved from Wikipedia: https://en.wikipedia.org/wiki/Optimism_bias

Wikipedia. (2020, September 07). *Task switching (psychology)*. Retrieved from Wikipedia: https://en.wikipedia.org/wiki/Task_switching_(psychology)#Switch_cost

About the Author

In this book, Alan draws on academic knowledge, industry best practice, and experience in the industry to explain how risks can be best managed in an ICT project.

Alan's academic past includes studying Data Science Probability at Harvard, Business Strategy at Wharton, Financial Risk at Imperial College London Business School, Business Administration at Wilfrid Laurier University, and Business Administration at McMaster University.

Alan also brings in-depth knowledge of international industry best practices. Alan maintains several of the most highly regarded formal certifications in project management including: Project Management Institute – Project Management Professional (PMI-PMP), Project Management Institute – Risk Management Professional (PMI-RMP), Axelos - Prince2 Agile Practitioner, Axelos Managing Successful Programmes (MSP), and Axelos Management of Risk (M_o_R).

Alan's experience in successfully delivering high-risk ICT projects spans over two decades. He has successfully delivered programs and projects across multiple industries including banking, insurance, financial regulation, oil & gas, mining, government, and national intelligence.

The projects Alan has delivered include relatively modest ICT projects ranging up to enterprise-wide ICT projects with budgets over $450 million.

Alan has managed project risks for blue-chip companies such as Deutsche Bank, JP Morgan Chase, Credit Suisse, Commerzbank, State Street Bank, Barclays Bank, The Prudential Insurance, The Financial Services Authority, Norwich Union/Aviva Insurance, ANZ Bank, Queensland Treasury, BHP Billiton Mitsubishi Alliance, Queensland Department of Education, Royal Dutch Shell, Santos, Brisbane City Council, Origin Energy, and the Australian Criminal Intelligence Commission.

Alan is no stranger to risk in his personal life. He has embraced and managed risk in adventure travel and adventure sport. Alan has embarked on a solo motorcycle adventure across Eurasia. He rode up through India to the Himalayas, across the world's highest motorable road, across Eurasia, and up to the Arctic Circle in Scandinavia. Alan also took his bicycle on a five-month solo journey across Africa. Alan has gone paragliding, scuba diving, and kiteboarding in some of the most exotic and remote locations across the globe. Having adventure-travelled through more than 75 countries, Alan is accustomed to identifying, assessing, and managing risks of all types.

Made in United States
Orlando, FL
06 November 2026

90276105R00136